Lecture Notes in Computer Science 1148

Edited by G. Goos, J. Hartmanis and J. van Leeuwen

Advisory Board: W. Brauer D. Gries J. Stoer

Springer

Berlin
Heidelberg
New York
Barcelona
Budapest
Hong Kong
London
Milan
Paris
Santa Clara
Singapore
Tokyo

Ming C. Lin   Dinesh Manocha (Eds.)

Applied Computational Geometry

Towards Geometric Engineering

FCRC'96 Workshop, WACG'96
Philadelphia, PA, May 27-28, 1996
Selected Papers

 Springer

Series Editors

Gerhard Goos, Karlsruhe University, Germany

Juris Hartmanis, Cornell University, NY, USA

Jan van Leeuwen, Utrecht University, The Netherlands

Volume Editors

Ming C. Lin
Mathematical and Computer Science Division, U.S.Army Research Office
Research Triangle Park, NC 27709-2211, USA
and
Department of Computer Science, University of North Carolina
Chapel Hill, NC 27599-3175, USA

Dinesh Manocha
Department of Computer Science, University of North Carolina
Chapel Hill, NC 27599-3175, USA
E-mail: (lin/manocha)@cs.unc.edu

Cataloging-in-Publication data applied for

Die Deutsche Bibliothek - CIP-Einheitsaufnahme

Applied computational geometry : towards geometric
engineering ; selected papers / FCRC '96 workshop, WACG
'96, Philadelphia, PA, May 27 - 28, 1996. Ming C. Lin ; Dinesh
Manocha (ed.). - Berlin ; Heidelberg ; New York ; Barcelona ;
Budapest ; Hong Kong ; London ; Milan ; Paris ; Santa Clara ;
Singapore ; Tokyo : Springer, 1996
  (Lecture notes in computer science ; Vol. 1148)
  ISBN 3-540-61785-X
NE: Lin, Ming C. [Hrsg.]; WACG <1996, Philadelphia, Pa.>; FCRC
  <2, 1996, Philadelphia, Pa.>; GT

CR Subject Classification (1991): I.3.5, I.3, I.2.9-10, J.6, J.2, J.3

ISSN 0302-9743
ISBN 3-540-61785-X Springer-Verlag Berlin Heidelberg New York

© Springer-Verlag Berlin Heidelberg 1996
Printed in Germany

Typesetting: Camera-ready by author
SPIN 10513796   06/3142 – 5 4 3 2 1 0   Printed on acid-free paper

Preface

Computational geometry (CG), as a discipline, has been intended to provide algorithmic foundations and analytic tools for geometric problems encountered in many fields of science and engineering. These include computer graphics, solid modeling, robotics, manufacturing, computer vision, astrophysics, geographical information systems, fluid dynamics, computational biology, etc. However, despite the wealth and abundance of literature and research, the intended technology transfer has been slow and limited.

The core of computational geometry can be enriched by new problem domains. At the same time, exposure to various applications will help in making CG more directly relevant. Today, computational geometry is in transition. To narrow the gap between theory and practice, a number of workshops have been organized in the last few years. Continuing the trend, the First ACM Workshop on Applied Computational Geometry (WACG), held as part of the second *Federated Computing Research Conference* (FCRC'96), was intended to bring together theorists and practitioners in computational geometry and related application areas. The main objectives were to identify factors that hinder timely and effective technology transfer and to foster dialogue and collaboration among different communities.

This volume contains invited contributions, state-of-the-art reports, and 12 contributed papers presented at WACG. The contributed papers were selected from a total of 32 submissions on the basis of the quality of the results and their relevance to the theme of the workshop.

We would like to thank Chee Yap for proposing the idea of organizing such a workshop at FCRC'96 and the Computational Geometry Advisory Committee for providing suggestions in the early planning stage. We are grateful to the Program Committee members for reviewing all the submissions. We wish to acknowledge the sponsorship of ACM SIGACT and SIGGRAPH, and additional support from the U.S. Army Research Office and National Science Foundation. Finally, we would like to thank all the speakers, panelists, and authors who contributed to this workshop.

We hope that the effort in organizing this workshop will start a new trend of *geometric engineering* and encourage cross-fertilization among different communities which share the use of geometric algorithms and techniques for applications in sciences, engineering, and computing.

August 1996

Ming C. Lin
Dinesh Manocha

Steering Committee

General Chair

Ming C. Lin

U.S. Army Research Office &
University of North Carolina, Chapel Hill

Program Chair

Dinesh Manocha

University of North Carolina, Chapel Hill

Program Committee

David Dobkin	*Princeton University*
Leonidas Guibas	*Stanford University*
Joseph S. B. Mitchell	*SUNY at Stony Brook*
Chee Yap	*New York University*

Advisory Committee

Herbert Edelsbrunner	*University of Illinois at Urbana*
Michael Goodrich	*Johns Hopkins University*
Leonidas Guibas	*Stanford University*
Kurt Mehlhorn	*Max-Planck-Institut für Informatik*
Joseph S. B. Mitchell	*SUNY at Stony Brook*
Emo Welzl	*Freie Universitat*
Chee Yap	*New York University*

Contents

Submitted Contributions

How Solid Is Solid Modeling?

Christoph M. Hoffmann

Department of Computer Science, Purdue University

1 On the Semantics of CSG and BRep

Constructive Solid Geometry (CSG) and Boundary Representations (Brep) are two major approaches to representing rigid solids dating back to the 1970s; see, e.g., [2, 6, 11, 14, 18, 22, 20, 21].

CSG implicitly represents a solid as an algebraic expression. The operators are regularized set operations, union, intersection and difference, and rigid-body motions. The operands are primitive solids, classically block, sphere, cylinder, cone and torus, instantiated to specific dimensions.

Brep explicitly represents the boundary of a solid as a data structure. Topologically, the surface is a quilt of vertices, edges and faces, where the adjacencies are represented. Geometrically, a face is a (well-behaved) subset of a surface. The surface could be a parametric surface or patch, an implicit algebraic surface, or a procedurally represented surface. The face boundaries are recursively represented as lower-dimensional boundary representations. Operations on solids in Brep first were the operations from CSG. However, soon other operations were introduced and implemented.

From the outset, research sought to give precise mathematical foundations to these solid representations and the operations on them. A proposed criterion by which to judge the representations was *informational completeness*. This was meant that there should be an algorithm that could, in principle, decide unambiguously whether any point in 3-space was inside, on the boundary of, or outside a given solid. Semantic work in CSG also paid attention to a finitary condition imposed to exclude pathological solids, for instance solids with a fractal boundary. The operations of (regularized) union, intersection, and difference were then defined in mathematical terms.

The task for defining a mathematical semantics for CSG and its modeling operations was simplified by the algebraic structure of the representation. The parallel task of giving Brep modeling a precise semantics turned out more difficult. Some research efforts formalized the topological validity of the representation, see, e.g., [18]. The interaction between topology and geometry, however, is a subject that continues to attract research; e.g., [10, 17, 23]

Over time, new operations were introduced into solid modeling that were difficult to fit into the well-established semantic framework of CSG. For example, consider a cube on which we round some edges and vertices. Conceptually, we can think of the construction as beginning with the cube, and modifying the shape by performing rounding operations. To accomplish this using only cubes, cylinders and spheres along with the Boolean operations of union, intersection and difference is not natural, hence a "rounding" operation was introduced in solid modelers.

The importance of such new operations to applications, and the apparent difficulty of reducing them conveniently to the repertoire of classical CSG is one of the factors that contributed, over time, to the decline of pure CSG modeling. While the conceptual legacy of CSG is very much present in many of today's solid modeling systems, Breps are used in virtually all of them. Moreover, the introduction of new operations has accelerated while needed semantic foundations are underdeveloped or absent. Two example areas follow that illustrate the situation, one with the classical operation of blending, the other with emerging design practices that stress ease of editing designs.

2   The Semantics of Blending

Rounding a convex edge or vertex, filleting a concave edge or vertex, are operations that are collectively called *blending* operations. Their precise semantics has a geometric and a topological aspect.

Geometrically, a blending surface is a surface that has to be in tangential contact with two or more given surfaces, along prescribed link curves, and whose geometric shape should conform to qualitative expectations. The geometric problem has been isolated and treated with precision by many researchers; see, e.g., [9, 11]. A blending surface qualitatively should have a rounded shape, and for this reason, spherically-derived blending surfaces are frequently chosen in practice. There are different ways to give meaning to the vague term spherically-derived. For example, in the case of blending two surfaces the following possibilities exist:

1. The blending surface is the envelope of the volume swept by a sphere that maintains simultaneous contact with the two primary surfaces; e.g. [12]. The *spine* of such surfaces is the trajectory of the center of the sphere, and it must be defined correctly in order to define the blending surface unambiguously.

2. A circle is swept in space such that, at each point, contact with the primary surfaces is maintained, thus defining a surface. Here, an addtional difficulty is to prescribe the spatial orientation of the circle as its center moves along a suitable curve in 3-space.

3. The surfaces to be blended are systematically deformed and pairwise intersected. The intersection curves lie on a blending surface. The circular rule might be that the intersection curves must pass through a fixed circle, in 3-space or in an abstract space; see, e.g., [13].

Other approaches, used for parametric primary surfaces, are derived from properties of the parametric representations; see, e.g., [9]. Figure 1 shows a constant-radius rolling-ball blending surface between two cylinders.

Consider variable-radius rolling-ball blends as a specific example. Such surfaces are obtained by rolling a ball whose diamater varies along the path, and using the surface of the swept volume as in the case of constant-radius rolling-ball blends. The major difficulty is to define precisely how the sphere's diameter

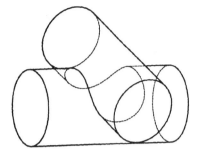 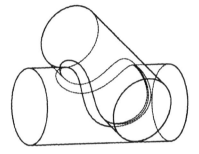

Fig. 1. Two cylinders and a rolling-ball blend between them

varies. The algorithmic techniques proposed in [19] are insufficient because they include an iterative step that traces an unspecified path on a two-dimensional manifold.

In the case of constant-radius rolling ball blends, the contact requirements imply that the spine of the blend must maintain a distance to the primary surfaces equal to the ball's radius. Thus, the spine is the intersection of the offset surfaces, of the primary surfaces, by the radius of the ball. Evidently, this is a necessary condition. Note, however, that it does not exclude global self-intersections of the blend.

In the case of variable-radius blends, we must determine the spine curve on the *equi-distance surface* of the primary surfaces, a generalization of the Voronoi cell boundaries familiar from computational geometry; e.g. [12]. The equidistance surface of two given surfaces f and g is the locus of all points p that are at equal distance from both f and g. Only in simple situations are such surfaces represntable with traditional mathematics. For example, the equidistance surface of a sphere and a (nonintersecting) plane is a paraboloid of revolution; the equidistance surface of two parallel planes is another plane; and the equidistance surface of two nonintersecting cylinders of equal radius and skew axes is a hyperbolic paraboloid. In general, the representaiton and analysis of equidistance surfaces requires higher-dimensional manifolds and projections; [7, 8].

While there is considerable work on the geometry of blending surfaces, work on the global topology and requirements for blending solid models is largely absent; [3]. A requirement of solid modeling is that blending surfaces be constructed based on the selection of edges and vertices, plus a few attributes. From this information, the blending operations proceed unassisted and in ways that are not even mapped out conceptually. For example, are the surfaces to be constructed sequentially or simultaneously? If they are to be serialized, in what order? For example, the two variants shown in Figure 2 have been obtained by blending the two edges in different order. A simultaneous blend might create a third variant. For a wide variety of such questions and how they might be answered see [3].

Clearly, such decisions, largely made automatically, affect the possible result of blending. Complete algorithms for blending of solid surfaces are of great practical importance, but there is little published work addressing the questions of topology and ordering of the possible variants.

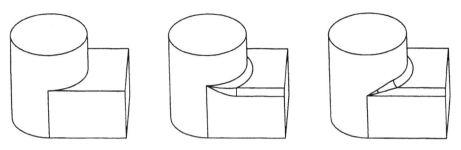

Fig. 2. Two variants when blending two adjacent edges

3 The Problem of Variational Design

Many solid modeling systems implement a design paradigm in which no longer
an specific solid shape is designed, but a parameterized class of potential shapes
in which a specific shape is then instanced. Figure 3 illustrates this for two-
dimensional shapes: A quadrilateral with a rounded corner has been defined.

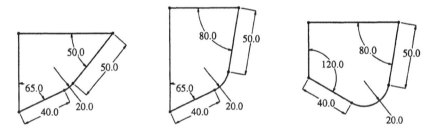

Fig. 3. Design variants of a quadrilateral

Some of the lengths and angles have been prescribed, as well as that two sides
should be perpendicular and the rounding arc tangent to the two adjacent edges.
By varying the lenghts or angles, different rounded quadrilaterals are obtained
from the same underlying definition schema. Moreover, this definition is descriptive rather than procedural because there are no requirements on the sequence
in which the constraints should be elaborated. This descriptive definitional style
is often termed *variational* in the solid modeling literature, and the procedural
definitional style is called *parametric*.

A basic semantic difficulty of variational design is that there are different ways
to interpret the design constraints and the meaning of the changes. In the case of
the rounded quadrilateral, the problem is the ambiguity of the constraints. For
example, the two interpretations shown side-by-side in Figure 4 are both correct

mathematically, but presumably one interpretation is intended while the other one is not.

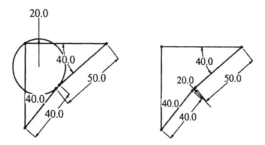

Fig. 4. Two interpretations of a design variant

We could argue that the multiplicity of interpretating variational geometric constraints is unavoidable. After all, the constraints will naturally correspond to a nonlinear system of equations, and such systems have multiple solutions in general. Hence, the question of "which variant" might be reformulated instead as the question of "what additional information" is needed, to unambiguously define the members of a class of designs in such a way that only the meaningful shape instances are included. Some mathematical and combinatorial difficulties arise in the investigation of this formulation; e.g., [1].

Another broad class of ambiguities in how to interpret a change in design arises from the persistent naming problem; [15, 16, 4, 5]. Consider Figure 5. Here, the part on the right was obtained by changing the position of the center of

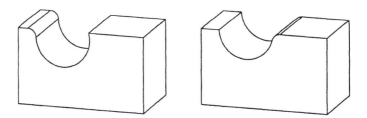

Fig. 5. Variant designs?

the round slot. This variation is counter-intuitive, because we probably consider such design changes conceptually as a continuous deformation of one shape into the other, here by raising or lowering the center line of the slot.

To understand why a solid modeling system could construct the variant we must consider the steps in which the part was designed: First, a block was created, with prescribed height, width, and depth. The block's shape could be varied subsequently by changing these dimensions. Next, the round slot was created by choosing a direction of the cut, a radius for the circular profile, and a position for the center, based perhaps on the distance from one of the sides and the bottom of the block. Finally, an edge round was created by selecting one of the edges bounding the slot laterally and prescribing a radius for the round. The edge selection was done visually, interacting with a specific instance.

When the slot's position is changed, we must recreate the data structure of the old instance to reflect the change. In particular, the design is reverted to the block, the slot is re-created from the new dimensions, and a new round is created. This means that a description of the edge has to be given that is independent of the data structure of the first design instance. The description has to be re-evaluated for the new design.

How could the edge be described? Clearly, it is the intersection of the slot with the top of the block — but so is the other edge, and it is this ambiguity whose resolution is imperfect in the example. Various schemata have been proposed in the cited literature. Moreover, proprietary schemata exist in implemented commercial solid modeling systems. These schemata are difficult to characterize.

The combination of the naming schema, and its reinterpretation after editing a design, constitutes a procedural semantics of variational solid design that is not clearly understood and requires further exploration. The difficulty of re-interpreting includes arbitrating multiple occurrences of a named entity, for example when subdividing an edge or a face, resolving clashes when merging vertices, edges or faces, and responding to the diappearance of named entities such as the obliteration of a vertex by an expanded feature elsewhere on the solid.

Consider the design shown in Figure 6. A protrusion was added to the block, but the height was chosen such that the top face, f_2, merged with the top of the block, f_1. The height of the prism is controlled by a dimension measured agains the top face. When the protrusion is lowered, a decision must be made which part of the L-shaped top face is to be used to reconstruct the prism. For a more detailed discussion of these and other problems see, e.g., [15, 16, 4, 5].

4 Summary

We have sketched several operations and design paradigms in solid modeling that remain without adequate semantic foundation. To obtain a proper mathematical semantics, and to harmonize it with the intuitions and expectations applications of solid modeling require, poses some fascinating and difficult research topics worthy of sustained exploration.

Acknowledgements Hoffmann gratefully acknowledges the support of the Office of Naval Research under contracts N00014-90-J-1599 and N00014-96-1-XXXX, and of the National Science Foundation under grants CCR 95-05745 and CDA 92-23502.

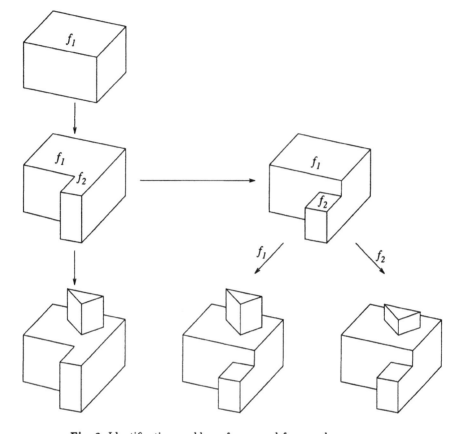

Fig. 6. Identification problems for merged faces and consequences

References

1. W. Bouma, I. Fudos, C. Hoffmann, J. Cai, and R. Paige. A geometric constraint solver. *Computer Aided Design*, 27:487–501, 1995.
2. I. Braid. The synthesis of solids bounded by many faces. *Comm. ACM*, 18:209–216, 1975.
3. Ian Braid. Nonlocal blending of boundary models. *Computer-Aided Design*, 1996. to appear.
4. V. Capoyleas, X. Chen, and C. M. Hoffmann. Generic naming in generative, constraint-based design. *to appear in CAD*, 1994.
5. X. Chen and C. Hoffmann. On editability of feature based design. *Computer Aided Design*, 27:905–914, 1995.
6. H. Chiyokura. *Solid Modeling with Designbase*. Addison-Wesley, 1988.
7. J.-H. Chuang. *Surface Approximations in Geometric Modeling*. PhD thesis, Purdue University, Computer Science, 1990.
8. J.-H. Chuang and C. Hoffmann. Curvature computations on surfaces in n-space. *Mathematical Modeling and Numerical Analysis*, 26:95–112, 1990.

9. G. Farin. *Curves and Surfaces for Computer-Aided Geometric Design*. Academic Press, 1988.

10. H. Hansen and N. Christensen. A model for n-dimensional boundary topology. In *Proc. 2nd ACM Symp on Solid Modeling and Applic.*, pages 65–73. ACM Press, 1993.

11. C. Hoffmann. *Geometric and Solid Modeling*. Morgan Kaufmann, 1989.

12. C. Hoffmann. Algebraic and numerical techniques for offsets and blends. In S. Micchelli M. Gasca, W. Dahmen, editor, *Computations of Curves and Surfaces*, pages 499–528. Kluwer Academic, 1990.

13. C. Hoffmann and J. Hopcroft. The potential method for blending surfaces and corners. In G. Farin, editor, *Geometric Modeling*, pages 347–365. SIAM, 1987.

14. C. M. Hoffmann and J. Rossignac. A road map to solid modeling. *IEEE Trans. Visualization and Comp. Graphics*, 2:3–10, 1996.

15. J. Kripac. *Topological ID system – A Mechanism for Persistently Naming Topological Entities in History-based Parametric Solid Models*. PhD thesis, Czech Technical University, Prague, 1993.

16. J. Kripac. A mechanism for persistently naming topological entities in history-based parametric solid models. In *Proc 3rd Symp on Solid Modeling*. ACM Press, 1995.

17. H. Lopes. *Algorithms for Building and Unbuilding Manifolds in Dimension 2 and 3*. PhD thesis, Catholic University of Rio de Janeiro, Dept. of Mathematics, 1996.

18. M. Mäntylä. *An Introduction to Solid Modeling*. Computer Science Press, 1988.

19. J. Pegna. *Variable Sweep Geometric Modeling*. PhD thesis, Stanford University, Mechanical Engineering, 1988.

20. A. Requicha. Mathematical models of rigid solids. Technical Report PAP Tech Memo 28, University of Rochester, 1977.

21. A. Requicha. Solid modeling - a 1988 update. In B. Ravani, editor, *CAD Based Programming for Sensory Robots*, pages 3–22. Springer Verlag, New York, 1988.

22. A. Requicha and H. Voelcker. Constructive solid geometry. Technical Report Tech. Memo 25, University of Rochester, Production Automation Project, 1977.

23. V. Shapiro. *Representations of Semialgebraic Sets in Finite Algebras Generated by Space Decompositions*. PhD thesis, Cornell University, Sibley School of Mechanical Engineering, 1991.

Robustness Issues in Geometric Algorithms

Steven Fortune

AT&T Bell Laboratories
Murray Hill, NJ 07974
sjf@research.att.com

1 Introduction

Geometric algorithms are usually described assuming that input data is in general position and that exact real arithmetic provides reliable geometric primitives. Often an implementer substitutes floating-point arithmetic for real arithmetic and uses real-world data, which might be degenerate by accident or design. Hence, the correctness proof of the mathematical algorithm does not extend to the program, and the program can fail on seemingly appropriate input data. This is the well-known problem of "robustness" in geometric computing[11].

This note[1] surveys recent research and current trends on the issues of numerical precision and degeneracies, addressing the assumptions of exact real arithmetic and general position, respectively. The references are impressionistic rather than complete, with the emphasis on recent rather than older work. Other surveys of robustness issues are also available [6, 11, 15].

2 Numerical precision

A geometric predicate is determined by the sign of an arithmetic expression. Sign-evaluation is exact in the conceptual model of the real numbers. However, a computer implementation cannot use real numbers. There are two scientifically plausible substitutions for exact real arithmetic. One is to substitute floating-point arithmetic for real arithmetic, and somehow deal with the resulting rounding error. The other is to substitute exact arithmetic on a subset of the reals, such as the integers or the rationals.

Floating-point arithmetic. Floating-point arithmetic is the default choice for much scientific programming. Used naively, it does not provide reliable geometric predicates. If an instance of a predicate is nearly degenerate, then the value of the corresponding expression can be very small, less than the rounding error in the floating-point evaluation of the expression. Hence the sign of the computed expression may well be erroneous. Usually it is possible to argue that the computed expression sign is the true sign for slightly perturbed coordinate data. Since coordinate data may well be imprecise originally, the erroneous sign may appear

---

[1] This note was adapted from a chapter of the report "Computational geometry: perspectives and challenges", edited by Bernard Chazelle.

to be innocuous. The difficulty arises with multiple predicate evaluations; there is no guarantee that any single global perturbation produces all the computed predicate values. Indeed, the computed predicate values may be geometrically inconsistent. Catastrophic implementation failure can easily result.

There are two broad categories of methods to deal consistently with floating-point rounding error. One category formalizes the notion of tolerances [5, 12]. A typical strategy might associate inner and outer tolerances with an object, say a point. If the inner tolerances of two points intersect, they are deemed coincident and merged; if the outer tolerances are disjoint, they are deemed separate; if neither case holds, the situation is ambiguous. As long as no ambiguity results during a computation, the result is correct. The hope is that ambiguities arise infrequently; an obvious drawback of this strategy is that is not clear what to do if an ambiguity does arise. Another drawback is that the generalization to complex geometric objects is not straightforward.

A second floating-point approach is modeled on the error analysis of numerical methods, particularly linear algebra. The goal is to show that a suitably implemented algorithm provides an answer that is in some precise sense near the mathematically correct answer. Error analysis of geometric algorithms requires consideration of both combinatorial and numeric structure. Often it is easy to argue that an algorithm produces combinatorially valid output[18], at least with suitably relaxed requirements. It has turned out to be much more difficult to argue that the numeric error associated with combinatorial structure is small. Full error analysis has been carried only for a few simple algorithms [8, 15].

Exact computation. Exact geometric computation requires that every predicate evaluation be correct [19]. This can be achieved either by computing every numeric value exactly, or by some symbolic or implicit numeric representation that allows predicate values to be computed exactly. Exact computation is theoretically possible whenever all the numeric values are algebraic, which is the case for most current problems in computational geometry.

With exact geometric computation, it is no longer reasonable to assume that each arithmetic operation takes constant time, as is the case with floating-point arithmetic. Rather, the cost of an arithmetic operation depends upon its context. Simple geometric predicates can be expressed as the sign-evaluation of an integer polynomial. The required arithmetic bit-length can be estimated from the degree of the polynomial and the bit-length of coordinates. For many predicates involving linear objects (such as orientation predicates) the degree is small, and the required bit-length is relatively minimal. However, even in simple cases software arithmetic is required, say over the integers or rationals, with a resulting increase in performance cost.

The cost of arithmetic also increases because of geometric constructions. A geometric constructor produces a new geometric object from old objects. Typically the coordinates of the new object can be expressed as polynomials in the coordinates of the old objects, and hence bit-length can be estimated from polynomial degree. For example, the coordinates of the intersection point of three planes have bit-length about three times the plane coefficient bit-length; the coefficients of a plane through three such points have bit-length about nine times the original plane coefficient bit-lengths. Hence an algorithm that uses geometric objects constructed to arbitrary depth can require arithmetic with prodigious bit-length, even if the algorithmic predicates are relatively simple.

Complex primitives on linear objects, or simple primitives on curved objects, apparently require that arithmetic be augmented with square roots, and, more generally, arbitrary polynomial roots. Such arithmetic operations can be implemented symbolically using general algebraic techniques such as resultants, Sturm sequences, and root separation bounds.

Discussion. Exact arithmetic is the most general technique that can guarantee the numerical reliablity of a geometric implementation. It appears not to be used widely, perhaps because of perceived performance cost, unfamiliarity, or lack of appropriate software tools.

There has been considerable recent progress in reducing the performance cost of exact arithmetic, at least for primitives with relatively minimal bit-length requirements. A useful strategy is adaptive-precision arithmetic. The predicate expression is first evaluated with low-precision arithmetic, perhaps with floating point, and only reevaluated at higher precision if the uncertainty bound exceeds the magnitude of the computed value. This strategy admits a rich spectrum of implementations [7, 13, 14, 17] with further exploration possible. Alternatively, there has been some progress at reducing the bit-length required for evaluation of simple predicates [3].

High-level geometric rounding algorithms may be able to address the coordinate growth that results from geometric constructions. Consider a polyhedron in three dimensions, with vertices, edges, and faces specified both by incidence structure and by coordinates. An exact computation may result in coordinates with large bit-length. An application may be satisfied with a short bit-length approximation to the numeric coordinates, as long as the combinatorial and numeric data are consistent. Rounding the coordinates of a polyhedron is not straightforward, however, since its combinatorial structure may be invalidated by small perturbations of its faces or vertices. A few examples [9, 10] demonstrate the complexity that results when high-level rounding is considered an integral part of a geometric algorithm; much more research needs to be done.

Efficient exact computation of geometric predicates involving algebraic numbers is just beginning to be explored [1, 14, 16, 19]. Many predicates that are well-understood in the linear domain (e.g. orientation) become much more complex in the curved domain, and their best evaluation using exact arithmetic is not well understood. Classical algebraic techniques can be used to give bit-length estimates required for expression evaluation; in many cases the bounds are dauntingly large [20]. It is possible that such bit-length estimates are too pessimistic, either in theory because the underlying algebraic machinery is not developed enough to give sharp estimates, or in practice because instances requiring long bit-length are infrequent.

3 Degeneracies

Degeneracies arise from the special position of two geometric objects. For example, two segments in general position either do not intersect or intersect at a point interior to both segments. Two intersecting segments in special position may overlap, may share a common endpoint with or without being collinear, may have one segment endpoint interior to the other segment, etc. Real-world data is likely to be degenerate. For example, segment endpoints may be explicitly chosen from a coarse grid, to facilitate interactive design.

The effect of degeneracy is to increase the number of special cases. While a sorting algorithm must deal only with the possibility of two keys being equal, a typical geometric algorithm faces the possibility of dozens or hundreds of different special cases. Since the overall correctness of an implementation may depend upon the correct treatment of special cases, the handling of special cases can permeate the implementation. This raises the obvious reliability concern that all cases have been considered and correctly handled.

Symbolic perturbation schemes [4] allow degeneracies to be resolved automatically. Conceptually, each geometric coordinate c_i is replaced with a symbolically perturbed coordinate $c_i + f_i(\epsilon)$, where $\epsilon > 0$ is unknown but very small and the perturbation function f_i is simple, say a polynomial. Substitution of the symbolically perturbed coordinates in a predicate expression results in a polynomial in ϵ with coefficients determined by the original geometric coordinates. The sign of the expression is given by the sign of the first nonzero coefficient, with coefficients taken in order of increasing powers of ϵ. For many classes of predicates, the f_i can be chosen to resolve all degeneracies.

Discussion. Symbolic perturbation is the subject of some controversy [2]. While it is certainly a useful tool in the implementation of geometric algorithms, existing schemes are not as applicable as might be desired. First, symbolic perturbation requires exact arithmetic, since the correctness of the perturbation depends upon exact evaluation of arithmetic expressions. Second, symbolic perturbation has been worked out in detail for only a small class of predicates. For example, constructed objects are often disallowed, since the perturbation function for a constructed object depends upon how the object was constructed and is much more complicated than the perturbation function for a primitive object. Perhaps most fundamentally, in a degenerate situation an algorithm implemented using symbolic perturbation does not solve the problem instance, but an arbitrarily-chosen nearby problem instance. This might be inappropriate in some applications, e.g, the symbolically-perturbed union of two polyhedra with abutting faces may not even be connected, if the perturbation pushes the faces apart.

4 Conclusions

Robustness has received considerable attention as a research area in the last few years, and many useful techniques have emerged. A sophisticated programmer can use the techniques to obtain an efficient implementation of a geometric algorithm that is as reliable as the implementation of a purely combinatorial algorithm.

However, the field is not mature, and the the application of the techniques is by no means routine. For example, a large application may well have both a kernel that requires complete reliability and outer layers that are less demanding. Thus the boolean operations of a solid modeler might require exact arithmetic to guarantee topological consistency, while the graphical rendering of a resulting solid can be done with less precision. Isolating a kernel that can be implemented robustly may be a significant design challenge.

References

1. C. Burnikel, K. Mehlhorn, S. Schirra, How to compute the Voronoi diagram of line segments: theoretical and experimental results. *Proc. 2nd Eur. Symp. Alg. (ESA 94)*, 1994.
2. C. Burnikel, K. Mehlhorn, S. Schirra, On degeneracy in geometric computations, *Proc. Fifth Annual Symp. Discrete Algorithms* pp. 16–23, 1994.
3. K. L. Clarkson, Safe and effective determinant evaluation, *33th Symp. on Found. Comp. Sci.* 387–395, 1992.
4. H. Edelsbrunner, E. Mücke. Simulation of simplicity: a technique to cope with degenerate cases in geometric algorithms. *ACM Trans. Graphics* 9(1):66-104, 1990.
5. S. Fang, B. Bruderlin, X. Zhu, Robustness in solid modelling – a tolerance based, intuitionistic approach, *Computer Aided Design*, 25:9, 1993.
6. S. Fortune, Progress in computational geometry, in *Directions in Geometric Computing*, Ch. 3, pp. 81–128, R. Martin, ed. Information Geometers Ltd, 1993.
7. S. Fortune, C. Van Wyk, Static analysis yields efficient exact integer arithmetic for computational geometry, to appear, *Transactions on Graphics*. See also Efficient exact arithmetic for computational geometry, *Proc. Ninth Ann. Symp. Comp. Geom*, pp. 163–172, 1993.
8. S. Fortune, Numerical stability of algorithms for 2d Delaunay triangulations, *International Journal of Computational Geometry and Applications*, 5(1,2), 193–213, 1995.
9. S. Fortune, Polyhedral modelling with exact arithmetic, *Proc. Third Symp. Solid Modeling and Applications*, pp. 225-234, 1995.
10. L. Guibas, D. Marimont, Rounding arrangements dynamically, *Proc. Eleventh Ann. Symp. Comp. Geom*, pp. 190–199.
11. C. Hoffmann, The problems of accuracy and robustness in geometric computation. *Computer* 22:31-42 (1989).
12. D.J. Jackson, Boundary representation modelling with local tolerances, *Proc. Third Symp. on Solid Modeling and Applications*, pp. 247–254 (1995).
13. P. Jaillon, Proposition d'une arithmétique rationnelle paresseuse et d'un outil d'aide à la saisie d'objets en synthèse d'images, Thèse, Ecole Nationale Superieure des Mines de Saint-Etienne, 1993.
14. S. Näher, The LEDA user manual, Version 3.1, January 16, 1995. LEDA is available by anonymous FTP from ftp.mpi-sb.mpg.de in directory /pub/LEDA.
15. Victor Milenkovic, Verifiable implementations of geometric algorithms using finite precision arithmetic. *Artificial Intelligence*, 37:377-401, 1988.

16. A. Rege, J. Canny, Fast point location for two- and three-dimesional real algebraic geometry, to appear, 1995.

17. J. R. Shewchuk, Robust adaptive floating-point geometric predicates, *Proc. 12th Ann. Symp. Comp. Geom*, pp. 141–150.

18. K. Sugihara, M. Iri, Construction of the Voronoi diagram for one million generators in single precision arithmetic, *First Can. Conf. Comp. Geom.*, 1989.

19. C. Yap, T. Dubé, The exact computation paradigm, 452–492, *Computing in Euclidean geometry*, D.Z. Du, F. Hwang, eds, World Scientific, 1995, second edition.

20. J. Yu, Exact arithmetic solid modeling, Ph.D. Thesis, Purdue University, 1992, available as CSD-TR-92-037.

Implementing Geometric Algorithms Robustly

Leonidas J. Guibas

Department of Computer Science
Stanford University
Stanford, Calif. 94305
e-mail: guibas@cs.stanford.edu

Abstract. This note is meant as a sequel to Steven Fortune's note on 'Robustness issues in geometric algorithms' [For96] in these proceedings. We revisit some of the issues raised by him, such as the consistency between the combinatorial and numerical data in geometric algorithms, and then we elaborate on a number of additional topics, including issues in proving correct geometric algorithms meant to be executed with imprecise primitives, and in the rounding of geometric structures so that all their features are exactly representable.

1 Introduction

Geometric algorithms are unique in that they operate on a mixture of numerical and combinatorial data. A typical geometric object, such as a convex polytope in E^3, involves both types of data. On the one hand the vertices of the polytope are points in E^3 and they are given via their coordinates (numerical data). On the other hand the facet structure of the polytope is essentially an embedded graph and is represented by a discrete structure (combinatorial data). Furthermore, the two types of data are inter-related. For the polytope to be convex, certain determinants defined by quadruplets of neighboring vertices in the graph structure must all have consistent signs. An algorithm manipulating a convex polytope exploits these inter-relationships for efficiency; and it has the responsibility of maintaining them in any new convex polytope it produces.

As Fortune points out, geometric algorithms cannot be implemented in the real RAM (random access memory) model in which they are normally developed by theoreticians. Real RAM machines do not exist — we have to either use inexact but speedy floating-point arithmetic, or exact but slow arbitrary precision integer or rational arithmetic. And, of course, the latter is not adequate for geometric problems where non-rational operations (e.g., length computations) are used. Though there have been a number of successful attempts to use mixed-mode arithmetic as a way to take advantage of the best features of each type [FV93], in order to have practical implementations of geometric algorithms on current hardware we have to live with either imprecise geometric primitives, or with rounding operations that reduce the bit complexity of derived geometric objects. Each of these limitations interacts adversely with keeping the numerical and combinatorial data of a geometric object consistent, as we now proceed to discuss.

2 Implementing algorithms with imprecise primitives

Very few geometric algorithms can be implemented as straight-line programs. The conditional tests found in geometric algorithms almost always involve computing the sign of an algebraic expression based on some of the numerical data to the problem — in fact, in a large number of cases the expression involved is a determinant [For93]. Thus numerical calculations are required not only in computing derived numerical geometric data (e.g., the intersection point of two lines in E^2), but also in numerical tests used for determining combinatorial structure (e.g., testing if a triangle defined by three points in E^2 is positively oriented, negatively oriented, or degenerate). Imprecision in the former type of calculation means that the derived numerical data may be slightly wrong; for the latter type of calculation it means that we may get a combinatorial structure which is no longer consistent with the actual numerical data of the problem. This type of sign error is often catastrophic — and is a common reason why straightforward implementations of theoretical geometric algorithms fail in practice.

Numerical analysis has developed an arsenal of methods, such as forwards and backwards error analysis, for deriving bounds on the errors in a numerical calculation, given some bounds on the errors of the primitives used. In a forward analysis we give error bounds on the computed data. In a backward analysis we give perturbation bounds on the input data, so that the computed data could be the result of an exact algorithm operating on some perturbed inputs. Unfortunately these method are not always applicable in the geometric context, since we do not have a satisfactory way to say when a combinatorial structure is 'near', or 'a small perturbation' of, another. For those problems where the input is purely numerical, backwards analysis makes the most sense. Fortunately, several common geometric problems fall in this category, for example the classic problem of computing the convex hull of a set of n points (say in E^2).

The output of a convex hull algorithm is purely combinatorial. In E^2, it is a circularly-ordered list of a subset of the given points. Given some error ϵ on the numerical primitives used by the algorithm (typically counterclockwise or 'sign' tests on triplets of the points and coordinate comparisons), we may hope to be able to give a (hopefully small) perturbation bound $f(n, \epsilon)$ on the input points and guarantee that the hull our algorithm computes is the real convex hull of the original points, each perturbed by no more than $f(n, \epsilon)$. Now most classical convex hull algorithms do not satisfy this condition. To see the difficulty, consider the easier problem of sorting n numbers. Suppose that our comparison routine is imprecise, and more specifically say that it can report '$a \geq b$', even though $a < b$, as long as $|a - b| < eps$. Now imagine we use an algorithm such as INSERTION-SORT [CLR90] to sort an array $A[1..n]$ whose elements are in reverse order ($A[1] > A[2] > \ldots > A[n]$), but any consecutive pair is very close ($|A[i] - A[i + 1]| < \epsilon$). It is conceivable that our imprecise comparison primitive will check the ordering of successive elements and report $A[1] \leq A[2] \leq \ldots \leq A[n]$, thus concluding that the array is actually sorted (when in fact $A[1]$ is much larger than $A[n]$). To paraphrase the way that Fortune [For96] says it, what is going on here is that even though for any consecutive pair of array elements $A[i]$, $A[i+1]$ there is a small perturbation of them that will make the comparison $A[i] \leq A[i+1]$ true, there is no *global* small perturbation that will make *all* the inequalities $A[1] \leq A[2] \leq \ldots \leq A[n]$ true at once. To say it a different way, 'greater than' is a transitive relation, but 'approximately greater than' is not!

A convex hull algorithm can get exactly into the same kind of difficulty. In the picture below it can process the points in (true) x-order and verify that each consecutive triplet is clockwise convex via an imprecise sign primitive, yet the resulting cycle can be arbitrarily far from being convex.

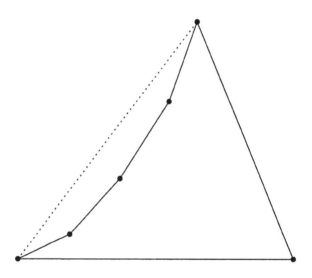

Fig. 1. A wrong convex hull

In order to have provably precise (at least in this backwards analysis sense) geometric algorithms we need to rethink the very mathematical structure via which the algorithm computes the desired result and ascertains its correctness. Unfortunately no mechanical transformation from a theoretical to a robust algorithm is known. Guibas, Salasin, and Stolfi developed the framework of *Epsilon Geometry* to address exactly this problem [GSS89] and were able to apply it to the convex hull problem. Fortune, Milenkovic, and others [For95, FM91, Mil88] also developed methods to address these issues for specific geometric problems (convex hulls, Delaunay triangulations, arrangements). But it is fair to say that we still lack a general theory on how to design geometric algorithms so that they behave robustly when implemented with imprecise primitives.

Several other difficulties also deserve to be mentioned. Take again the convex hull problem in E^2 we just discussed. Should we really be satisfied with an output which is the convex hull of our inputs slightly perturbed? Because of the perturbation involved, the hull we actually get back from our algorithm can be unsatisfactory in one of two ways. Firstly, some points in our set might actually be slightly outside the 'convex hull' — perhaps this type of error we can live with. But secondly, it can be that our 'convex hull' is not really a convex polygon, but only approximately so; in other words, some of its angles can actually be slightly concave. Now this error is more serious, as typically the output of a convex hull algorithm is fed to other algorithms as input, and these algorithms do need real convexity of their input polygon for their correct execution. If this situation pertains, then we must guarantee that our computed convex hull is truly convex. This gives rise to the requirement of 'strong convexity' and make the convex hull algorithm design problem much more challenging [GSS90].

The difficulty with imprecise primitives mentioned above is like that of 'drift-errors' in numerical computation. Locally each primitive can be made correct by a small perturbation of its inputs, but there is no global small perturbation which makes all the primitives valid simultaneously. An even more serious difficulty arises because the primitives used often share input data, and thus are not mathematically independent. In other words, once we select signs for certain conditional tests, the signs of other tests may mathematically follow, yet numerically we may have the latter test come out the wrong way. The result is that the execution path our algorithm follows is unrealizable for any real geometric data.

For an example of this, consider the classical algorithm of producing an arrangement of lines in the plane incrementally, adding the lines one after the other. When a new line arrives, we trace it through the faces of the arrangement defined by the prior lines. A series of orientation tests, dual to the counterclockwise test for points mentioned above, is used to compute how the new line exits the current face it is in. Classical geometric theorems, such as a version of Pappus's theorem from projective geometry [Ped70], may determine how the new line must exit the current face dependent upon how it exited earlier faces. If we make the wrong exiting decision we will end up computing an arrangement whose topology may be unrealizable by any set of real lines. One the other hand, to really compute which of our orientation tests are implied by earlier such tests is a very tough problem; it is equivalent to deciding the truth of a statement in the difficult existential theory of the reals [Mne89]. The only way out seems to be to allow an answer that is an arrangement of 'approximate lines', or pseudolines. We will take this up again in the following section.

3   Rounding geometric structures

Perhaps all the above difficulties will make us want to reconsider exact arithmetic, despite its high computational costs. Nevertheless, as we remarked earlier, integer or rational arithmetic is not always sufficient for solving a geometric problem. A closer analysis reveals another difficulty: the bit-complexity of the numbers we need is roughly additive in the number of operations used to construct each of our derived geometric objects. Thus deeply nested constructions are expensive, and the more we operate on previously computed data, the more expensive we make operations on them. The only way out of this tar-pit seems to

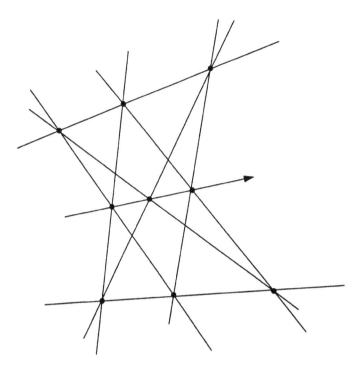

Fig. 2. Pappus's theorem

be to *round* our geometric data every so often, i.e., to replace them with others requiring less precision (equivalently, having smaller bit complexity).

Now rounding numbers is well understood, but when we talk about rounding geometric structures we run up against that familiar enemy: keeping the numerical and the combinatorial data of a problem consistent. To go back to our earlier convex polytope example, we must allow rounding to collapse nearby vertices and thus possibly make some vertex determinants zero, but we sure hope that the rounding will still keep the polytope convex. What this means is that we cannot round each vertex independently — instead we must use a global rounding algorithm which guarantees that the signs of the original vertex tetrahedra certifying convexity may become zero, but do not invert. How to do this is non-trivial.

In general rounding is trying to accomplish two somewhat opposing tasks. We naturally want rounded geometric features to stay as close to their original counterparts to within the allowed precision/resolution. But we also want to maintain all the consistency constraints between the numerical and symbolic parts of the structure. To see better what this means, consider again the above arrangement example. Say we would like each vertex of our arrangement to be mapped to a point of certain fixed precision — to be specific let's say an integral point (a point with integer coordinates). We cannot just round each vertex to its nearest integral point, as this might cause a vertex to pass through an edge and change the topology of the arrangements. Of course we must allow some features to collapse because of the rounding: a vertex may be made to lie on an edge, or all the vertices defining a tiny face of the arrangement may come to the same integral point. But we do not want to allow a vertex to pass through an edge. This kind of 'topological inversion' violates our consistency requirements between the numerical and combinatorial data of the problem. As we remarked earlier, to accomplish such a rounding of a line arrangement, we must allow each original line to be replaced by a nearby polygonal line. But in fact we may have to break up a line into more pieces than those necessitated by the arrangement vertices lying on it, in order to maintain topological consistency.

How do to this has been addressed in several papers. The most economical of these roundings (in the sense of breaking up the lines as little as possible) is the so-called *snap-rounding* of Greene and Hobby, an example of which is shown below for the case of line segments [GM95]. See also [GY86, Mil90, FM91].

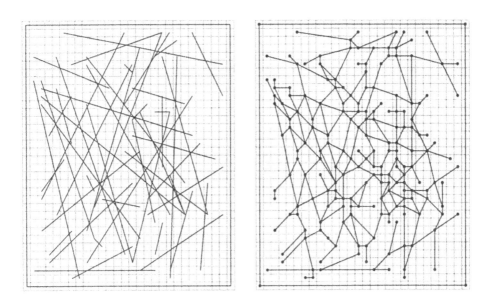

Fig. 3. A line segment arrangement and its snap-rounded form

As in the previous section, there several additional challenges to be overcome here. Another highly desirable property of such a rounding is that it be *canonical*, in the sense that it is a function of set of original objects and the allowed resolution only, and not of the history of the structure. In other words, if we add a line to a rounded arrangement and round the result, then delete that line and round the result, we would like to be back exactly where we started. This is not easy to accomplish, unless each rounded structure keeps around references to the original data from which it was created. In [GM95] it is shown how to accomplish this for the line-segment arrangement problem, in the dynamic context of segment insertions and deletions.

A seemingly unattractive feature of such rounding schemes is a *phase dependence* on the grid location: if we shift our grid of integral points slightly, we may get a different and non-congruent rounding of the arrangement. But it turns out that this variability is essential to guaranteeing topological consistency, while having original arrangement features either collapse or stay some minimum distance apart.

4 Conclusions

We hope that the discussions and examples of the previous sections have shown the deep and subtle issues involved in implementing geometric algorithms robustly. Although significant progress on this problem has been made, it is clear that much more remains to be done in developing both the theoretical foundations of how to keep combinatorial and numerical data consistent, and in fully understating the software engineering issues of how to organize geometry libraries so that floating point or rounding errors can be made as innocuous as possible.

References

[CLR90] T. H. Cormen, C. E. Leiserson, and R. L. Rivest. *Introduction to Algorithms*. The MIT Press, Cambridge, Mass., 1990.

[FM91] S. Fortune and V. Milenkovic. Numerical stability of algorithms for line arrangements. In *Proc. 7th Annu. ACM Sympos. Comput. Geom.*, pages 334–341, 1991.

[For93] S. Fortune. Computational geometry. In R. Martin, editor, *Directions in Computational Geometry*. Information Geometers, 1993. To appear.

[For95] S. Fortune. Numerical stability of algorithms for 2-d Delaunay triangulations. *Internat. J. Comput. Geom. Appl.*, 5(1):193–213, 1995.

[For96] S. Fortune. Robustness issues in geometric algorithms. In *Proc. 1996 Workshop on Applied Computational Geometry*, page in these proceedings, 1996.

[FV93] S. Fortune and C. J. Van Wyk. Efficient exact arithmetic for computational geometry. In *Proc. 9th Annu. ACM Sympos. Comput. Geom.*, pages 163–172, 1993.

[GM95] Leonidas Guibas and David Marimont. Rounding arrangements dynamically. In *Proc. 11th Annu. ACM Sympos. Comput. Geom.*, pages 190–199, 1995.

[GSS89] L. J. Guibas, D. Salesin, and J. Stolfi. Epsilon geometry: building robust algorithms from imprecise computations. In *Proc. 5th Annu. ACM Sympos. Comput. Geom.*, pages 208–217, 1989.

[GSS90] L. Guibas, D. Salesin, and J. Stolfi. Constructing strongly convex approximate hulls with inaccurate primitives. In *Proc. 1st Annu. SIGAL Internat.*

Sympos. Algorithms, volume 450 of *Lecture Notes in Computer Science*, pages 261–270. Springer-Verlag, 1990.

[GY86]   D. H. Greene and F. F. Yao. Finite-resolution computational geometry. In *Proc. 27th Annu. IEEE Sympos. Found. Comput. Sci.*, pages 143–152, 1986.

[Mil88]   V. J. Milenkovic. Verifiable implementations of geometric algorithms using finite precision arithmetic. *Artif. Intell.*, 37:377–401, 1988.

[Mil90]   V. Milenkovic. Rounding face lattices in d dimensions. In *Proc. 2nd Canad. Conf. Comput. Geom.*, pages 40–45, 1990.

[Mne89]   N.E. Mnev. The universality thorems on the classification problem of configuration varieties and convex polytope varieties. In O.Y. Viro, editor, *Topology and Geometry – Rohlin Seminar*, pages 527–544. Springer Verlag, 1989.

[Ped70]   D. Pedoe. *Geometry, a comprehensive course*. Dover Publications, New York, 1970.

Robustness in Geometric Algorithms

Franco P. Preparata

Department of Computer Science
Brown University
115 Waterman, Providence, RI 02912-1910.
email: franco@cs.brown.edu

Issues of robustness have become central to the development of geometric algorithms in recent years. The original assumption of a computational model obtained by extending the traditional RAM to real-number arithmetic proved less innocent than originally thought. To equate floating-point arithmetic to real-number arithmetic turned out to be indefensible on the basis of failures in practical applications. Another convenient assumption has been the hypothesis of "general position," which dispenses with the detailed consideration of special cases. Unfortunately, degenerate conditions (colinearity, cocircularity) which are likely to be generated by coarse-grid data as they occur in practice, give rise to numerically critical events. Failures originating from these assumptions have fundamentally hindered the adoption of computational geometry by practitioners.

Over the years several approaches have been proposed to remedy these shortcomings. An excellent and succint survey of the underlying philosophies is represented by the introductory note prepared for these proceedings by Steven Fortune (see also by Steven Fortune, "Computational geometry: perspectives and challenges," B. Chazelle, editor).

With reference to the effects of approximate arithmetic, there is speculation that no single approach may be capable of conferring robustness to geometric algorithms. Presumably, several tools may be included in an arsenal designed to achieve robust computations.

However, it is not entirely out of the question that there be approaches that may evolve into the closest thing to a methodology. The cautious tone of the previous statement is suggested by the experience that a methodology for robustness has been a rather elusive goal.

Such sufficiently general approach could derive from the following observations . The numerical computations of a geometric algorithm are basically of two types: tests (predicates) and constructions. These two types of computations have clearly distinct roles. Tests are associated with branching decisions in the algorithm that determine the flow of control, whereas constructions are needed to produce the output data of the algorithm.

Input data (whether the result of empirical observations or not) are reasonably assumed to be expressed with b bits, for some small integer b. Approximations in the execution of constructions give rise to approximate results, which may be not only acceptable but also mandated: indeed, as long as their maximum absolute error does not exceed the resolution required by the application (such as spacing of raster lines in graphics), it would be wasteful to produce substantially more accurate results.

On the other hand, approximations in the evaluation of predicates may produce an incorrect branching of the algorithm. Such event may have catastrophic consequences, giving rise to *structurally* incorrect results. Therefore, tests are much more critical, and their execution must be carried out with complete accuracy. We shall therefore characterize geometric algorithms not only on the basis of their performance, but also on the basis of the complexity of their predicate computations. In a large class of geometric algorithms, tests are based on the evaluation of the sign of a multivariate polynomial.

The preceding considerations elicit to the conclusion that tests must be carried out with complete accuracy, whereas some tolerance is permitted for constructions. (It must be observed, however, that such tolerance should not lead to contradict the topological structure of the result as provided by the tests.)

Complete accuracy seems to revert us back to the infinite precision implied by real-number arithmetic. Fortunately, the inherently coarse nature of the input data comes to the rescue in this connection. Each predicate is expressible as the sign of a multivariate polynomial in the input variables. The degree of this polynomial is a measure of the complexity of the predicate (since for degree d, exact evaluation may have to performed with $db + O(1)$ bits, conventionally assuming input variables of degree 1), and the maximum degree of the predicates evaluated by an algorithm becomes the sought additional criterion of simplicity.

Of course, the deployment of an arithmetic engine corresponding to the maximum precision is not prescribed each time a predicate is evaluated. Such power is called for only in near-degenerate cases. In typical cases, much lower precision may be sufficient to confidently evaluate the predicate. This leads to the notion of *adaptive precision*, where the arithmetic engine consists of a sequence of *evaluators* of increasing power and cost, each paired with its *certifier*, a device designed to guarantee the validity of the evaluator's result. The design of such (evaluator, certifier) pairs is a major challenge in this approach to robustness.

Finally, the arithmetic degree of an algorithm is not only an *a posteriori* measure of its performance, but should be used as an (*a priori*) design criterion in the development and adoption of low degree primitives.

Only extensive experimentation with a variety of test cases can substantiate the feasibility of the outlined approach.

Applications of Computational Geometry in Mechanical Engineering Design and Manufacture

Michael J. Pratt

Rensselaer Polytechnic Institute
Center for Advanced Technology
and
National Institute of Standards and Technology
Manufacturing Systems Integration Division.
E-mail pratt@cme.nist.gov

Abstract. This contribution has a threefold purpose. Firstly, it urges a wider interpretation of 'Computational Geometry' than is often taken. Secondly, it emphasizes the importance of the choice of representations as well as algorithms in geometric computing. Thirdly, it provides examples from the author's experience, supporting the opinions éxpressed.

1 Introduction

It appears that the term 'computational geometry' was first used by Robin Forrest. Certainly he published a paper with that title in 1971 [4], and another in 1974 [5]. Forrest used the term primarily in connection with the study of free-form curves and surfaces as used in the design of such artefacts as aircraft, cars, ships, shoes and plastic mouldings. His emphasis was primarily on representational issues, not algorithms. The present writer co-authored a book, published in 1979 [3], taking the same view of computational geometry. It can be argued that this is a restricted view, and that there are many other areas of application of geometric computing. But it has to be said that the application of free-form geometry for practical purposes requires such an enormous volume of numerical data that it would be impossible without the use of computers — this, I believe, is why Forrest initially coined the name.

The seminal computational geometry text by Preparata and Shamos appeared in 1985 [13]. With its emphasis on the complexity of algorithms for the solution of problems mainly involving 'simple' geometric elements such as points, lines and planes, this view of the field is also a very restricted one. However, for some reason a large body of researchers adopted the term 'computational geometry' to denote this kind of activity alone. The free-form curve and surface enthusiasts then had to be content with some other name for their research area, and chose 'Computer Aided Geometric Design' (CAGD). This is an unfortunate choice of terminology, since the work of the CAGD community is primarily mathematical, and bears very little relation to design in any true sense of the word.

Forrest's original focus was on representational issues in complex geometry; the Preparata/Shamos focus was on algorithmic issues involving simple geometry, albeit often in complex situations. Both areas of work, in my opinion, are merely subfields of a larger field to which the name 'Computational Geometry' rightly belongs. Whenever I use the term in the remainder of this paper it should be interpreted in the wider sense.

I am convinced that computational geometers should devote as much attention to representations as they do to algorithms, particularly in dealing with applications relating to engineering design and manufacture. For one thing, some forms of geometric representation are totally inadequate for engineering purposes. One just does not see polyhedral cars on the road, for example, nor polyhedral aircraft in the skies. The F117 stealth fighter and B2 bomber may be cited as exceptions — but even these are not convex!

If we accept that representation is important, then the following points apply:

1. The choice of representation strongly influences the choice of algorithm.
2. 'Simple' representations may give rise to clearly defined algorithms amenable to complexity analysis.
3. Complex representations are required for many engineering purposes, but the associated algorithms are often themselves complex and not very amenable to analysis.
4. In such cases, improvements can only be determined in measured rather than analytical terms.

Before going on I will expand briefl on the third of these points. Consider an algorithm for computing the intersection curve of two free-form surfaces, perhaps represented in Bézier or NURBS form. At the outset, the problem exhibits two levels of geometric complexity. At the local level it presents singularities such as self-intersections, cusps and self-tangencies, while at the global level the algorithm may have to detect and keep track of multiple disjoint branches of the intersection curve [12]. Suppose that a 'marching' algorithm is used. In this case it is first necessary to determine a starting point on each branch, and then to step along the branch from this point to compute further points along its length. The step-length may need to vary according (for example) to estimates of surface curvature generated as the computation proceeds. Special measures may need to be invoked to handle singularities if any are detected. Some means must be found for stopping the process once a closed loop has been traversed and the starting point passed (it is unlikely that this point will be hit exactly on completion of the loop, and multiple traversals of closed intersection loops is known to be a problem with at least one commercial CAD system).

I believe that it is not possible to provide a useful analysis of the overall complexity of this kind of algorithm, though individual portions of it may be separately analysed. But in any case the performance of the algorithm will be strongly dependent on the pair of surfaces it is given to intersect. The only practical way of assessing its performance is therefore to test it on a large number of cases and to compare its performance on a statistical basis with that of other algorithms or earlier versions of the same algorithm.

2 Cutter Path Determination in NC Machining

The computation of cutter paths for numerically controlled (NC) machining is significantly more complex even than the surface intersection algorithm just described. Several different cases arise, depending on the machining strategy and the nature of the surfaces involved. One particular strategy is discussed below in some detail.

In the machining of free-form parametric surfaces there are advantages in the use of cutters with composite profiles, such as radiused end-mill or toroidal cutters. These have cylindrical sides and flat ends with a blending radius between the two. Optimal cutting conditions for metal removal result when the tool contact point with the surface being machined lies on the toroidal blending region of the tool. It is clear that, given the tool contact path on the surface, the computation of points on the path of the tool reference point (the centre of the flat end) is not a simple matter. It is as bad or worse for other cutter geometries which are sometimes used, including barrel-shaped or conical (in both these examples the sides of the tool are the cutting surfaces). The more complex tool geometries are used to best advantage in the context of what is known as 4 or 5-axis machining. In 3-axis machining the tool may be moved simultaneously in the x, y and z senses but the direction of its axis remains constant. In 4 or 5-axis machining the direction of the tool axis is also under continuous control. The advantages of this complex type of machining strategy are that (i) a much finer finish can be obtained on the machined surface, which obviates the need for expensive manual finishing with hand-held grinders, and (ii) optimal cutting conditions, as described above, may be maintained where the tool contacts the surface being machined. The disadvantage, it will readily be appreciated, is the much greater mathematical complexity of the necessary cutter path calculations. Another disadvantage of the use of more general cutter geometries is that while the tool may contact the surface at the desired point on the designated cutter path, it may also interfere with the surface in an undesired manner at some other point, causing gouging. The automatic detection of such situations adds further complication to an already very computation-intensive procedure.

Consideration must also be given to the way in which the surface region to be machined is bounded. It may, in fact, be surrounded by other surface regions, each of which also has to be machined, and it is clearly important that in machining the first surface the cutter should not gouge any of the others. Accordingly, there is a requirement for computing the locus of tool contact points on one surface such that the tool is simultaneously in contact with (but not gouging) a second surface.

In practice, the computation aims to determine a sequence of tool positions such that the tool is not actually in contact with both surfaces, but rather is within some predefined tolerance of both of them. The algorithm first computes what are known as *common normals*. These are two vectors, each normal to one of the surfaces and also to the surface of the cutting tool. The determination of these vectors requires a complex iterative procedure. Once the vectors are known, their cross-product gives a first approximation to the direction of tool motion, and a notional step is taken in this direction, the step length being determined by some estimate of the curvature of the tool path. The resulting position of the tool will in general leave it out of tolerance with both surfaces, but a further iterative process may be used to bring it back within tolerance. The cycle is now repeated for the determination of the next step in the tool path. This is necessarily a very abridged description of the calculation process; further details are given by Faux and Pratt [3].

Despite their complexity, it is worth noting that algorithms of the kind described have been implemented for many years in practical systems routinely used by many manufacturing companies throughout the world.

3 Case Studies in Representational Issues

In the following subsections I will present several case studies, some dating from the last decade, to illustrate the importance of representational issues in geometry for engineering design and manufacture. I hope that they will serve to illustrate some of the points made in the Introduction.

3.1 Case Study 1 — Planar Polyhedral vs. Curved Surface Modeling

This Case Study describes solid modelling experiences in the mid-1980s. At that time three approaches to solid modelling existed in commercial CAD systems, as follows (representative examples from the mid-1980s are given):

CSG: —
- NONAME (developed at Leeds University in England, and again available in commercial systems)
- PADL-2 (developed in the USA at the University of Rochester, and at one time incorporated into more than one commercial system)
- Synthavision (Developed and marketed by the Magi Corp., later acquired by CADAM Inc.)

Faceted B-rep: —
- CATIA (developed and marketed by Dassault Systèmes, France)
- Euclid (marketed by Matra-Datavision, France)
- Geomod (developed and marketed by SDRC, USA)

'Exact' B-rep: —
- Design (developed by MDSI in the USA, but suppressed when the Schlumberger Corp. acquired MDSI)
- Proren (developed at Ruhruniversität Bochum in Germany, and marketed by ISYKON GmbH)
- Romulus (developed and marketed by Shape Data, UK, currently owned by EDS/Unigraphics — its successor from the same source is Parasolid)

It must be emphasized that the situation has changed a great deal in the last decade. Some of the systems listed have disappeared from the market-place, and others have appeared in their place. There are now no commercially available CAD systems based on the pure CSG approach, and most of those described above as using the faceting approach now use exact geometry. At the time in question, CATIA was primarily a surface modeller, and was only just beginning to acquire a solid modelling capability.

Before proceeding further, an explanation of the fate of the CSG approach will be given. At first sight, this type of modelling has the advantages of robustness and theoretical elegance, but in its pure form it is now little used in engineering. Its major problems are

– Lack of flexible approaches to model-building
– Lack of explicit geometry in the primary model
– Difficulty of implementing free-form surface types such as NURBS

However, CSG has undoubtedly left a legacy in practical CAD systems in the wide availability of diverse procedural modeling methods. However, these are now usually implemented in a boundary representation context.

Moral 1: Robustness and elegance alone do not ensure survival!

However, the main purpose of this subsection is to discuss the faceted approach to modelling. Faceting was used by some developers to avoid the need, in Boolean and other modeling operations, to compute surface intersections other than plane/plane cases. At this time the computation of intersections between general pairs of surfaces still posed many problems, and facteing was a way of circumventing or putting off those problems. Of course there was a tradeoff — as model accuracy (and hence the number of facets) increased, the number of plane/plane intersections also increased. Ultimately it would have been 'cheaper' to compute the true curved surface intersections, however computer-intensive the necessary algorithms.

Faceted models are still widely used in applications such as computer graphics, where speed currently has higher priority than geometric accuracy, and in solid freeform fabrication (SFF). But solid modeling in the engineering context is very sensitive to geometric accuracy. The problems include

– Geometry/topology discrepancy
– Limiting case problems

These two difficulties are illustrated and discussed in the following subsections.

Geometry/topology discrepancy. For an illustration of geometry/topology problems arising from faceting, consider the case of two circular cylinders, tangent along a generator:

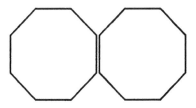

Fig. 1. Polyhedral Approximation of Two Cylinders

The faceted approximations have no contact. But if one is rotated by $2\pi/16$ they intersect along *two* lines. Either way, the modeler computes the wrong topology. In general, such situations can lead to disagreement between the geometry and topology in a model. In this case the modeller may become totally confused if an attempt is made to apply further modelling operations to an already erroneous model.

Another application area where faceted models are clearly inadequate is in the automated analysis of functional tolerances. This requires, in particular, the use of exact cylindrical surfaces.

A Limiting Case Example — the Blending Problem. In the 1980s, one commercially available faceting CAD system used the following method for blending sharp edges between two surfaces:

1. The user was prompted to select end points of the edge to be blended and to provide values of blend radii at these points.
2. The system constructed a trimmed surface to act as the blend surface.
3. An operation was then invoked to create a solid having the blend surface as one of its faces.
4. Integration of the blend with the original model then called for a Boolean union operation to unite the two solids. Although this system had some exact geometry capabilities, Boolean operations were always at this time performed on faceted or polyhedral approximations where objects with non-planar faces were concerned.

In my experience this method failed far more often than it worked. A happy choice of blend radii could usually be found, but this required extensive trial and error. A typical problem part on which this particular system failed was the 'Cranfield Object', illustrated in the figure below — this was a test object used to evaluate blending in solid modelers. It is, in fact, a real object some 4 metres high, part of an oil rig. The two versions shown (generated by different systems) both date from the mid-1980s.

Why then did the blending consistently fail? Consider the following points:

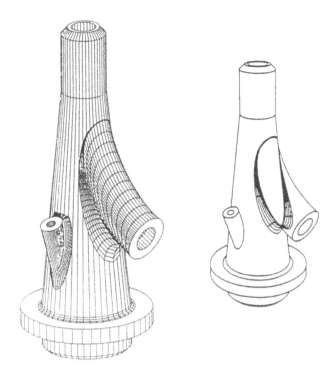

Fig. 2. The 'Cranfield Object'

1. A 'true' blend surface meets tangentially with the surfaces being blended.
2. For the Boolean operation, intersections between the blend surface and the two original surfaces had to be computed. In the 'exact' case the intersections will be curves of tangency.
3. However, the system tried to compute these curves in a faceting environment, by intersecting faceted surfaces.
4. Since these surfaces were approximately tangent, and each contained many facets, there was a very good chance that one pair of intersecting facets from each surface would be parallel, to within the modeler's internal geometric accuracy.
5. The plane/plane intersection algorithm (and hence the entire blending operation) in this case failed, causing the system to crash.

It must be emphasized that all this was more than ten years ago. The system concerned has changed dramatically during that time, and dispensed with the use of polyhedral approximations some years ago. Also, in common with most modern solid modelling systems, it now does not crash if a computational error occurs. Instead, having detected the error such systems roll back to the situation which existed before the most recent modelling operation was invoked, and inform the user. In some cases the problem will result from a user error; if this is not so, then it is usually possible to find an alternative means of attaining the desired end which avoids the computational problem.

'Exact' B-rep Modelers. It is interesting to note the progress made during the 1980s by the developers of 'exact geometry' systems. They tackled the surface intersection and blending problems head-on, from the beginning. Their systems were slow at first, but soon speeded up as algorithms and hardware improved.

The developers of faceting systems found it very difficult to change their ways; they took *about ten years* to catch up with the geometric capabilities of the exact systems.

Moral 2: Don't put off tackling problems that at first seem geometrically difficult, and don't oversimplify in the interests of easy computation.

3.2 Case Study 2 — Car Body Design

My next example is again concerned with representational issues. Car producers usually use off-the-shelf CAD systems these days. One exception is Company A in Germany, who use an in-house system. This generates mathematical surfaces that are C^2 (curvature) continuous. Company A says that this property is essential to the aesthetic appearance of its cars.

On the other hand, Company B in England uses a commercially available CAD system. They model their car body models in terms of surface patches that are not even C^1 (tangent-plane) continuous where they meet — Company B says that discrepancies of magnitude up to 0.5° in surface normal direction do not matter.

Why not? Well, once the shape is defined, NC cutter paths are generated, press dies are cut, and the die surfaces are hand finished. After the body panels are pressed, an inelastic deformation called springback occurs in them. Company B finds that all of these are smoothing processes, so that small geometric anomalies in the mathematical model leave no trace in the manufactured body panels.

The interesting question for me is 'Which company is right'? Both are successful companies, building sought-after cars. Maybe they are both right, and Company A really does have some good reason for requiring its mathematical model of the car body to have curvature continuity. For the reasons cited by Company B, however, there are likely to be significant deviations between this mathematical model and the actual shape of the manufactured car.

A further example is perhaps even more striking. Company C in the USA for many years used an internally developed surface design system (though they have recently given up its use in favour of a standard commercially available CAD system). I had some experience with the in-house system some years ago, and it became clear that some of the surfaces it generated actually had holes in them — narrow slits extending between opposite boundaries, their width reducing to zero at the boundary itself. However, since the graphical display was wire-frame, these slits did not show on the workstation screen! Furthermore, for most practical purposes the slits had no effect. For example, there was no problem in generating cutter paths on such surfaces for die machining — computed cutter paths are piecewise linear or piecewise circular, and the computational procedure just took a slightly larger step to jump across the gap! In fact a generation of successful cars was designed using this system, despite its possession of what at first sight might appear to be a fatal flaw.

Moral 3: What the eye doesn't see, the heart doesn't grieve over. Choice of representations and associated algorithms should only be made after careful analysis of the precise requirements of a new geometric application. Sometimes it is possible to get useful results from a representation that at first sight seems defective.

3.3   Case Study 3 — Verification of NC Machining

My third example concerns further work done in the 1980s, by what was then General Dynamics in San Diego [6], which much impressed me at the time. The objective was to verify the correctness of programs written to generate machined parts using numerically controlled (NC) milling machines. The verification method was as follows:

1. The initial stock material was modelled using the solid modeler TIPS [15].
2. For each (5-axis) cutting tool motion specified in the NC program, TIPS was then used to compute the swept volume. Models of the cutting tools were needed for this.
3. A Boolean subtraction of the swept tool volume from the stock material model was performed.
4. This procedure was repeated until all the tool motions had been dealt with.
5. The resulting volume was then compared with a model of the finished part as designed, to verify correctness of the NC program. Shaded graphical images were produced by the system.

At the time, this computation involved an awesome amount of number-crunching. The computation was performed on a dedicated IBM mainframe, which ran overnight. The TIPS modeler used (a public domain system) was extremely cumbersome and inefficient, but the results obtained were sufficiently useful to make the disadvantages bearable. Industry is often prepared to attain its computing goals quickly by using massive computer resources rather than more slowly by seeking the most efficient algorithms. Now, of course, with increases in hardware speed, such simulations and verifications as those described above can be run faster than real-time on modern workstations.

Moral 4: For engineering purposes, find something that works first, however slowly, and then seek improvments. Eventually, either the process will become fast enough for practical purposes on the original equipment, or improved hardware speed will make it fast enough on new equipment.

4 A Personal Note

In one sense this whole paper is personal, since it draws on experiences in modelling for CAD and CAM over the past twenty years or so. But before concluding I will say a few words about the directions in which I would like to see engineering applications of geometry develop. A growing number of people, including myself, are becoming interested in the possibility of replacing the polyhedral models used for many purposes by models having curved surfaces, represented implicitly. Some of the surface types which have been examined for this purpose are

1. Piecewise quadrics [7, 8]
2. Piecewise algebraic cubics [14]
3. Piecewise Dupin cyclides [9, 10]
4. Piecewise supercyclides [2, 11]

The advantages are that smoother and more realistic models will result, made up of far fewer patches than the polyhedral models currently used. I need hardly add that, from the point of view of a computational geometer in the wider sense, investigation of these possibilities will give rise to a large number of interesting representational and algorithmic problems!

5 Discussion and Conclusions

The examples presented earlier have been aimed at demonstrating the very complex nature of some of the geometric computations arising in engineering design and manufacturing, and also at emphasizing the importance of a good choice of representations in geometric computing. I list below several areas related to computer-aided design and manufacturing where research is currently in progress:

- 'Traditional' boundary representation modellers are designed to represent objects with 2-manifold boundaries. However, modern requirements demand, amongst other things, the modelling of ojects with internal structure, e.g. finite element meshes. This has led to research into appropriate data structures for what is loosely called *nonmanifold modeling.*
- Most CAD systems now have a means for the automatic generation of blended edges on grossly modelled shapes of engineering parts. The geometric aspects of this problem are reasonably well understood, but the topological problem of handling the many different ways in which designers might like these blends to interact where they meet is far from solved.

– Modern CAD systems represent not only geometry and topology, but also parametrized entities and geometric constraints. Currently there exists no 'canonical' representation for models of this form, and strenuous efforts are under way to find such a representation. The object is to find an abstract model such that the internal representations used by all existing CAD systems can be mapped onto it. The practical objective is to permit the exchange of parametrized, constraint-based models between different systems, a frequent industrial requirement.

– Different CAD systems use different types of representations, particularly for free-form surfaces such as those used in car body design. One French CAD system uses polynomial parametric surface patches of degree up to 20. Most other systems do not exceed 15 in the degree of their polynomials. It is often necessary for companies to exchange geometric data with other companies, for example subcontractors, who may use different systems. To import polynomial data of degree 20 into a system using maximum degree 15 requires the use of *degree reduction*. The resulting surface will in general be made up of a larger number of patches than the original, each patch being of lower degree, and the whole surface lying within some specified tolerance of the original.

– An important problem which has recently emerged from experiments in model exchange arises from differences in the numerical accuracy to which different solid modelling systems work internally. In general, there is no problem in transferring a model from a system with 'tight' numerical tolerances into a system with looser tolerances. However, a transfer in the opposite direction may well fail because elements in the transmitted model which are connected when judged by the standards of the sending system may not be connected according to the standards of the receiving system. In this case geometry/topology discrepancies arise, and the model must somehow be 'repaired' before it can be further worked on in the receiving system.

Many further research topics in this and other application areas are described in the recent Computation Geometry Impact Task Force Report by Chazelle et al. [1]. As regards lessons learned from past experience in the field of geometric computing in engineering, some of my personal findings may be summarized as follows:

– The most appropriate geometric representation must be chosen for the job. Determining this may not be a trivial exercise.

– Maximally efficient algorithms must be chosen, appropriate for that representation.

– It is important to develop something that performs the required task *first*, worrying about improving efficiency should come later — the initial effort provides a baseline for improvement.

– Don't worry if the algorithms are too complex to be analyzed — they often will be, in practical situations.

– But by all means concentrate on creating libraries of efficient algorithms which are frequently needed by more complex applications.

References

1. B. Chazelle et al., Application Challenges to Computational Geometry, Technical Report TR-521-96, Princeton University, Princeton, NJ (1996)
2. Degen, W. L. F., Nets with Plane Silhouettes II. To appear in Proc. IMA Conf. *Mathematics of Surfaces VII*, Dundee, Scotland, 2 – 4 September 1996. Oxford, England: Oxford University Press (in preparation)
3. Faux, I. D., Pratt, M. J.: *Computational Geometry for Design and Manufacture*. Chichester, England: Ellis Horwood (1979)
4. Forrest, A. R.: Computational Geometry. Proc. Roy. Soc. Lond. **A 321** (1971) 187 – 195
5. Forrest, A. R.: Computational Geometry — Achievements and Problems. In R. E. Barnhill and R. F. Riesenfeld (eds.) *Computer Aided Geometric Design*, Academic Press (1974)
6. Fridshal R., Cheng, K.P., Duncan, D., Zucker, W.: Numerical Control Part Program Verification System. In Proc. Conf. on CAD/CAM Technology in Mechanical Engineering, Massachusetts Institute of Technology, March 1982. MIT Press (1982)
7. Guo, B., Representation of Arbitrary Shapes using Implicit Quadrics. The Visual Computer **9** (1993) 267 – 277
8. Laporte, H., Nyiri, E., Froumentin, M., Chaillou, C., A Graphics System based on Quadrics. Computers and Graphics **19** (1995) 251 – 260
9. Martin, R. R., de Pont, J., Sharrock, T. J., Cyclide Surfaces in Computer Aided Design. In Gregory, J. A. (ed.) *The Mathematics of Surfaces*. Oxford, England: Oxford University Press (1986)
10. Peternell, M., Pottmann, H., Designing Rational Surfaces with Rational Offsets. Technical Report No. 28, Institut für Geometrie, Technische Universität Wien, Austria (1995)
11. Pratt, M. J., Quartic Supercyclides I: Basic Theory. To appear, Computer Aided Geometric Design
12. Pratt, M. J., Geisow, A. D., Surface/Surface Intersections. In Gregory, J. A. (ed.) *The Mathematics of Surfaces*. Oxford, England: Oxford University Press (1986)
13. Preparata, F. P., Shamos, M. I., Computational Geometry: An Introduction. New York: Springer-Verlag (1985)
14. Sederberg, T. W., Techniques for Cubic Algebraic Surfaces (Parts 1 and 2). IEEE Computer Graphics & Applications **10** (1990) 5, 14 – 25 and 6, 12 – 21
15. TIPS Working Group, *TIPS-1: Technical Information Processing System*. Institute of Precision Engineering, Hokkaido University, Japan (1978)

On Some Applications of Computational Geometry in Manufacturing and Virtual Environments

(Extended Abstract*)

Joseph S. B. Mitchell

Applied Mathematics and Statistics, SUNY Stony Brook, NY 11794-3600, USA;
jsbm@ams.sunysb.edu, http://ams.sunysb.edu/~jsbm/jsbm.html

Abstract. There are many motivating applications for the study of computational geometry problems. In this talk, we briefly touch on two such application domains — manufacturing and virtual environments. We discuss a few of the applications-driven computational geometry problems in these domains and point to some possible future research directions that will assist in transferring algorithmic methods into practice.

1 Manufacturing

Advanced manufacturing has been identified as one of the critical technologies for economic competitiveness. The goal is to make manufacturing systems more flexible and more automated, from the design stage, to the prototyping, process planning, and quality control stages. Flexibility and automation require sophisticated software tools and systems, often utilizing computationally intensive algorithms. Geometry arises in many, if not most, of the computational problems of manufacturing, due to the inherently geometrical nature of form design, of manufacturing processes, and of the physical products themselves.

Thus, it is natural that the field of *computational geometry*, which is dedicated to the algorithmic study of geometric problems, should be keenly involved in the development of new advanced manufacturing technologies. The motivation for many of the problems studied by computational geometers over the field's (roughly) two decades of existence came from needs arising in application domains that are certainly relevant to manufacturing — solid modeling, robotics, VLSI, computer graphics and visualization.

Manufacturing engineers deal with extensive computations involving geometry, using whatever methods they can find that work adequately in practice. Recent efforts by computational geometers have started to address some more formal (algorithmic) aspects of these manufacturing problems. In particular, some initial work has been done to understand the complexity of questions of

---

* Copies of most of the transparencies for the talk are available electronically at the author's web site.

manufacturability of parts by various processes, of design and tolerancing of CAD models, of layout of two-dimensional patterns from sheet stock, etc.

In the manufacturing process planning problem, the goal is to automatically generate a "good" (if not "optimal") process plan to manufacture a part, given its geometry, the geometry and material properties of the raw material, and the geometric and dynamic constraints of the process and tools. Instead of automatically generating a process plan, it is also important to have efficient tools for *verifying* the correctness of a given plan, which has been automatically or manually generated. There are many processes that require study and analysis, including numerically-controlled (NC) machining (e.g., pocket milling), gravity casting, injection molding, bending/folding, cutting, deposition/etching, and various forms of rapid prototyping (stereolithography, fused deposition modeling, laser sintering, etc.).

In the manufacturing design arena, there are also a host of problems that arise of a geometric and algorithmic nature. Important graphics problems related to CAD and modeling include surface and volume rendering. The automatic mesh generation problem has received a great deal of attention in both the numerical analysis community and the computational geometry community, due to its fundamental role in the finite element method and its natural use of so many structures from computational geometry (most notably Delaunay triangulations and octrees and their variants). Several recent papers in graphics and computational geometry have proposed methods of model simplification, using "decimation", refinement, wavelets, and polyhedral separation modelled as a covering problem. The optimal use of material (e.g., sheet stock) has received considerable attention in the garment, footwear, aircraft, paper, and furniture industries, where the goal is to utilize fabric, leather, sheet metal, cardboard, paper, and lumber in the most efficient way, minimizing waste. Layout and routing problems arise extensively in both small-scale (e.g., placement and wire routing VLSI) and large-scale (e.g., hydraulic tubing routing within aricraft and ship structures) manufacturing. It is important not just to design structures that are feasible, but also to design structures that are easily assembled (e.g., taking advantage of the geoemtry and compliance of parts to assist in the assembling task). Finally, in integrating systems together that must share the same space, or in virtual testing of servicability and manufacturability, it is important to be able to conduct rapid interference detection between geometric models. (We mention this problem again in the next section.)

While it is encouraging to see the results of early efforts in studying the algorithmic issues associated with manufacturing, much more work is needed before we can really say that the field of computational geometry has a major impact on manufacturing technology. It is especially important that there be a continuing and strengthening exchange of expertise and discussion between the various disciplines of manufacturing engineering and the computational geometry community.

Sources of Further Information

The role of computational geometry in manufacturing has received increasing recognition in both the computational geometry and manufacturing communities. There was an NSF-sponsored Workshop on Manufacturing and Computational Geometry, held at New York University on April 1-2, 1994, and organized by David Dobkin, Harry Stephanou, and Chee Yap. A report on the results of this workshop is widely available [8].

As a further side effect of the workshop, and evidence that the two communities realize the importance of coming together, a forthcoming special issue of *Algorithmica* (a leading journal in theoretical algorithmic research) is devoted to the interface between computational geometry and manufacturing, *and*, most recently, a call for papers went out for a special issue of the journal *Computer-Aided Design* (a leading journal in the applications of CAD to manufacturing), devoted to "Computational Geometry and Computer-Aided Design & Manufacturing", editted by R. Janardan and T. Woo.

The discussion of the role of computational geometry in manufacturing is continuing. An overview of many important issues is summarized in the recent Computational Geometry Impact Task Force Report [2]. Other important special issues and reports on manufacturing technology issues in computer science include [1, 6, 7].

2 Virtual Environments

In virtual environments, geometry also plays a central role. The term *"Virtual Reality"* (VR) has come to mean the use of computer hardware and software to simulate a model of a physical environment in such a way that humans can readily visualize, explore, and interact with the "objects" in the environment. Since physical environments are inherently geometric, many of the computational problems involved in designing and building a VR system are geometric in nature. The goal is to avoid constructing a physical mockup by using the virtual environment to take a virtual tour and check repairability/maintainability.

Two of the fundamental problems that arise in VR and are currently computational bottlenecks are (1) visualization and rendering of highly complex models of three-dimensional environments, at interactive frame rates, and (2) real-time interference detection between moving models in a virtual environment.

In some of our recent research at Stony Brook, in collaboration with the virtual reality group at Boeing, we have examined the problem of intersection detection: *Preprocess an environment, $\mathcal{E}$, into a compact data structure to support efficient queries: "Does object F intersect any of the obstacles in $\mathcal{E}$?"* There have been many methods devised to attack this problem, including those based on BSP trees, R-trees, octrees, Voronoi diagrams, sphere-trees, and, most recently, "BoxTrees" and closely related "OBB-trees". We have implemented and compared experimentally several methods, including one based on the notion of "meshes of low stabbing number" (to track the motion of a moving object within a tetrahedral mesh of free space that hopefully has the property that few tetrahedra are intersected by the moving object), basic box-based methods (simple grid of boxes, k-d trees, and variants of R-trees), and some more general "bounding volume trees" (BV-trees). We have found that BV-trees yield highly competitive performance and show promise of being the method of choice. See [3, 4, 5] for more information, as well as pointers to other references.

Several interesting implementation issues arise in attacking these problems, including

- We must, of course, make our algorithms robust.
- We must deal with data that is not only degenerate, but is also "bad" – surfaces self-intersect, have "cracks", duplicate faces.
- We must assemble effective and standardized datasets that reflect what one expects to see in practice. (This depends, of course, on the application domain!) We attempt also to generate "random" data that models (approximately) situations seen in practice.
- Almost all algorithms utilize many parameters — some values are preset, others are selectable at runtime — whose values can dramatically affect the *performance, robustness*, and (possibly) the *correctness* of the code. We must study how to select parameters automatically, using clues given by the geometry of the data.
- We must devise efficient methods for handling very large datasets that cannot fit in internal memory, so that our algorithms take into account the high cost of paging from secondary storage.

References

1. "Information Technology for Manufacturing: A Research Agenda," *National Academy Press*, 1995.
2. B. Chazelle, et al., "Application Challenges to Computational Geometry", Technical Report TR-521-96, Princeton University, April 1996; also available on the web at http://www.cs.princeton.edu/~chazelle/taskforce/CGreport.ps
3. M. Held, J. Klosowski, and J. S. B. Mitchell, "Evaluation of Collision Detection Methods for Virtual Reality Fly-Throughs", In proceedings *Seventh Canadian Conference on Computational Geometry*, Québec City, Québec, Canada, August 10-13, 1995, pp. 205–210. See http://ams.sunysb.edu/~jsbm/jsbm.html for full version.
4. J. Klosowski, M. Held, J.S.B. Mitchell, "Collision Detection for Fly-Throughs in Virtual Environments", *Proc. Twelfth Annual ACM Symposium on Computational Geometry*, Video Proce edings, 1996.
5. J. Klosowski, M. Held, J.S.B. Mitchell, "Real-Time Collision Detection for Motion Simulation within Complex Evironments", to appear, *SIGGRAPH'96*, Technical Notes.
6. Stucki, P., Bresenham, J., Earnshaw, R., Guest Editors. *Rapid prototyping technology*, Special issue of *IEEE Computer Graphics and Applications* 15 (1995), 17–55.
7. Wozny, M.J., Regli, W.C., Guest Editors. *Computer science in manufacturing*, Special issue of *Communications of the ACM* 39 (1996), 33–85.
8. C. K. Yap, "Manufacturing and Computational Geometry Workshop Report", *Computational Science and Engineering*, IEEE Computer Society Press, 1995, pages 82–84.

Visualizing Geometric Algorithms – State of the Art

David Dobkin*

Department of Computer Science
Princeton University
Princeton, NJ 08540
dpd@cs.princeton.edu

Abstract. The visual nature of geometry makes it a natural area where visualization can be an effective tool in communicating ideas. This is enhanced by the observation that much of the action in computational geometry occurs in 2 and 3 dimensions, where visualization is highly plausible. Given these observations, it is not surprising that there has been noticeable progress during the past few years in the production of visualizations of geometric algorithms and concepts. There is every reason to believe that this will continue and even accelerate in the future. In this note, I briefly survey the current state of the art as well as suggesting new directions that should be pursued in the future. Further details appear in the survey article [HD96].

As anyone who has tried to implement a complex geometric algorithm knows, implementing geometric algorithms is a difficult task. Conventional tools are limited as aids in this process. The programmer spends time with pen and pencil drawing the geometry and data structures the program is developing. This problem could be solved by the use of visualization tools. In the ideal world, this visualization would be used for 3 purposes; demonstration, debugging and isolation of degeneracies. Ideally, we would like to use the same tools for all 3 functions. So, we would use the tool to help us debug the implementation of an algorithm by providing visual interaction during the debugging process. Next, we would like to use the same tool to create a visualization of the algorithm with which the user can interact. This interaction can be either passive or active. For example, a video tape provides passive interaction since the viewer's controls are limited to the VCR controls. Active interactions allow the viewer greater control over the visualization as we describe further below. Finally, there is the issue of isolating problems in code that is symbolically correct. Typically, such bugs come from degeneracies either in the data or in the computational model. Visualization has the potential to be a great help here as the tool allowing the user to jump into the code at (or preferably before) the point at which it breaks.

---

* This work supported in part by the National Science Foundation under Grant Number CCR93-01254.

Of these 3 purposes, we have seen the most success in demonstration. After 5 years, the video proceedings of the computational geometry conference continue to thrive. Researchers in various computing environments are able to create videos of visualizations that are satisfying and useful adjuncts to understanding the algorithms or systems which they visualize. However, there is significant room for improvement. Many of the visualizations submitted to the review are prepared in ad hoc fashion. We are quite distant from having a set of tools that rivals LaTeX or even TeX in any form. Further, there is no agreement on the styles to be used to produce high quality visualizations. Computational geometry videos are different in nature from the recent high quality mathematical visualizations that have been produced at the Geometry Center (eg Not Knot [EG91] or Outside In [LM94]). Visualizations in computational geometry tend to make minimal use of high quality graphics preferring instead "low budget" visualizations. The objects being visualized support this decision. It makes little sense to use smooth rendering techniques to hide edges and vertices of a polyhedron when these are the defining components of the shape. Instead techniques are typically chosen which highlight these components. The videos that have been produced serve as excellent adjuncts to papers that have been or are about to be published. The challenge remains to develop tools that make the process easer. Ideally, we should expect that a paper describing a geometric algorithm will always be accompanied by an animation of the algorithm.

The problem of creating active interactions remains largely unsolved. It is still the case that a visualization demonstrates the behavior of an algorithm on one sample input and explains the behavior of the algorithm on that input. A better scenario would allow the to not only specify the input, but also to interact with the view (and possibly even the input data) as the algorithm is running. There are a few systems that exist that allow the user to interact with a running animation (e.g. [Tel], [TD95]).

However, the interactions come at a price. The viewer must typically have the hardware that was used to develop the interaction. This limits the ability to integrate such animations into hypermedia documents. There is hope that the emergence of Java and VRML will help remove this limitation.

The hope is that the existence of Java as a programming language and VRML as a file format will enable interactions to be easily transmitted over the World Wide Web. This is far from a solved problem. There remain various speed problems. Java does solve much of the bandwidth problem by doing a single download of the algorithm. However, there remains the speed of execution problem. The interpreted nature of Java may make interaction impossible for complex algorithms. Further, there is a need for toolkits on which Java applets can be built. Ideally, there would be a computational geometry set of classes that would achieve wide use in the community. These classes would make code sharing easier and would allow implementors of new algorithms to begin at a higher level. This hope also arises in the other uses of visualization.

Ideally, Java will not only enable interactions, but will also enable implementors to use their applet as it is being developed as an aid in debugging their implementation. The visual nature of geometry makes Java a natural tool for debugging. Rather than drawing by hand the emerging data structures and geometry, an applet that displays the situation would be a wonderful adjunct. Disregarding issues of speed, it is not difficult to create the plumbing necessary to allow a developing applet to run in tandem with a conventional debugger which provides control over the running process. This combination should prove useful to all implementors.

Beyond classes and tools that simplify implementation, implementors also need data sets to be used as inputs to algorithms. Various issues arise here. First there should be generators for simple data sets that can be used for demonstration. The simplest example is N random points or segments in the plane. Tools need to be created as in [Sch], [EKK...] to create and modify such sample inputs. These are absolutely necessary for simple debugging of implementations. Next, there need to be more subtle input sets (eg points in (or near degenerate situations). Finally, there need to be collections of data from real applications. It is unnecessary for each implementor to reinvent input sets. Indeed, we as a community must create a set of classes and sample inputs that are usable for a wide range of applications.

I have attempted to describe here the current state of visualization for geometric algorithms. To summarize, the field is doing well but there are exciting opportunities for the future. The video review has become well established within the field. It is now time to take a closer look at the possibilities of hypermedia. A first step in this direction is the construction of appropriate classes upon which Java applets can be built for animations. An accompanying task is the generation of good sets of test data on which algorithms can be implemented. The value of such work will lie both in easier communication of results within our community and in an increased ability to reach those outside of our community.

References

[EG91]   David Epstein, Charlie Gunn, *et al*, "Not-Knot", distributed by A K Peters, Wellesley MA, 1991.

[EKK...] P. Epstein, J. Kavanagh, A. Knight, J. May, T. Nguyen and J.-R. Sack, "Workbench for Computational Geometry" in *Animation of Geometric Algorithms: A Video Review*, edited by Brown and Hershberger, DEC SRC TR 87b, 1992.

[HD96]   A. Hausner and D. Dobkin, "Making Geometry Visible: An introduction to the Animation of Geometric Algori thms", in *Handbook for Computational Geometry*, edited by Sack and Urrutia, to appear.

[LM94]   Silvio Levy, Delle Maxwell and Tamara Munzner. "Outside In", distributed by A K Peters, Wellesley MA, 1994.

[Sch]   Peter Schorn, Adrian Brüngger and Michele De Lorenzi, "The XYZ GeoBench: Animation of Geometric Algorithms" in *Animation of Geometric Algorithms: A Video Review*, edited by Brown and Hershberger, DEC SRC TR 87b, 1992.

[TD95] A.Y. Tal and D.P. Dobkin, "Visualization of Geometric Algorithms", IEEE Transactions on Visualization and Computer Graphics (TVCG) Volume 1, Number 2.

[Tel]   Seth Teller, "Visualizing Fortune's Sweepline Algorithm for Planar Voronoi Diagrams" in *Symposium on Computational Geometry, 1993 Video Review*, edited by Brown and Hershberger.

Geometric Algorithm Visualization, Current Status and Future

D.T. Lee*

Department of Electrical and Computer Engineering
Northwestern University
Evanston, Illinois 60208, USA.
E-mail: dtlee@ece.nwu.edu

Abstract. We give a survey of the current status of geometric algorithm visualization and offer some suggestions regarding geometric software library and future directions for visualization software.

1 Introduction

Since its inception two decades ago computational geometry has become a very active research field within theoretical computer science. There are a good number of research publications collected in pub/geometry/geombib.tar.Z, available via anonymous **ftp** from ftp.cs.usask.ca. Several journals dedicated to computational geometry have been established. The reader is encouraged to visit the Web page on *Geometry in Action* by D. Eppstein at http://www.ics.uci.edu/~eppstein/geom.html and computational geometry page by J. Erickson at http://www.cs.berkeley.edu/~jeffe/compgeom.html for more information.

Only recently an informal assessment of the impact of the field on other science and engineering disciplines was conducted and the questions of its *relevance* to practice were raised among researchers within the community. See the task force report prepared by Chazelle *et al.*[3]. As a result of these discussions and the need of geometric software, activities concerning development and implementation of geometric algorithms ensued. The annual ACM Symposium on Computational Geometry started to have a video session and entertain communications of experimental results beginning in 1991. The first International Workshop on Computational Geometry Software sponsored by NSF and ONR was held at the Geometry Center, University of Minnesota in January 1995, and effort in collection of geometric software has been underway. See the Web page at http://www.geom.umn.edu/software/cglist.

In addition to the LEDA[10] project at the Max-Planck-Institut für Informatik, Saarbrücken, Germany, a concerted effort of building a kernel for geometric software has been initiated in Europe, known as the CGAL project[7].

* Supported by the Office of Naval Research under the Grants No. N00014-93-1-0272 and No. N00014-95-1-1007 and by the National Science Foundation under the Grant CCR-9309743.

There are other projects related to the efforts of building geometric software, including GASP[15], GeoLab[4], GeoMAMOS[8], XYZ Geobench[13], etc. In this article we'll concentrate on geometric algorithm visualization tools, give a summary of their current status and give a *wish list* of what a geometric algorithm visualization tool or environment ought to provide.

2 Current Status

Using conventional approaches it is difficult to understand the structures of geometric objects represented in numerical format. Perception of objects is better obtained from visualization, as graphical form is most natural to human eyes, and conveys more information than numerical values or texts. By the same token to better understand the behavior of an algorithm, be it geometric or not, we would make use of visualization. Techniques and tools for algorithm visualization are thus developed in hopes that they not only help describe the algorithm behavior after they are fully implemented, but also help algorithm developers debug the programs when they are being implemented. We describe below briefly projects related to geometric algorithm visualization. There are software tools for *program visualization* or general algorithm animation including Zeus developed by Brown[2], Tango[14] by Stasko and Pavane by Roman *et al.*[12]. The reader is referred to [11] and to a special issue on Visualization (*Computer* Vol. 27, No. 7, July 1994) for more information.

Alpha-shape[5] developed at the National Center for Supercomputing Applications, University of Illinois, is a special purpose *shape* modeler and visualization package. It takes a set of points in 3-D space and computes the **alpha**-shape defined by the point set for different values of α, a control parameter. The computed output for various values of α is first stored in binary form in a temporary data file, which is later used as input to a graphical visualizer. The user selects a specific value of α and the pre-computed **alpha**-shape gets displayed. The software is available via anonymous **ftp** from ftp://ftp.ncsa.uiuc.edu/Visualization/Alpha-shape.

GASP[15] developed at Princeton University is a tool for geometric algorithm animation. It has a library of primitive animation functions. The user stores the needed animation routines in a *style* file, and in the main application program the user will invoke these animation routines at appropriate places. All input and output operations are performed using files. The software written in C is available via anonymous **ftp** from ftp.cs.princeton.edu:/pub/people/ayt/gasp.tar.Z.

GeoLab[4], developed at the Universidade Estadual de Campinas is a programming environment for implementation, testing and animation of geometric algorithms. This tool uses shared libraries of algorithms and an incremental approach to aggregating new types of geometric objects, data structures and extensions accessed through dynamic linking. It runs on SparcStations under Sun/OS using

XView graphics library. It is written in C++ and has been used as a research and teaching tool at the Universidade Estadual de Campinas.

GeomView[9] developed at the Geometry Center is a general purpose data visualization software. It can be used as a stand-alone *viewer* for geometric objects or as a *display engine* for objects produced by other programs. Specifically it takes as input a geometric data file in a special format and displays the contents in graphic form. Various operations, such as translation, rotations, zoom-in and zoom-out are provided to let the user visualize and manipulate the graphical data. It runs on Silicon Graphics(SGI) IRIS workstations and NeXT workstations.

GeoSheet[8] developed at Northwestern University, is an interactive visualization tool designed to simplify geometric algorithm visualization procedures in a distributed environment. It is display device independent, although in the current release the display is based on Xfig (Facility for Interactive Generation of Figures under X11) and is primarily for 2-dimensional objects. An interprocess communication (IPC) mechanism is used for message passing between processes that may be running on different machines. This mechanism allows the user to visualize geometric objects without storing the data into a file. GeoSheet is implemented in C++ running under UNIX and X windows environment. The 3-dimensional version, **3Dsheet**, is under development and runs on SGI machine. The current version contains some geometric algorithm implementations using LEDA. See http://www.eecs.nwu.edu/~theory/geomamos.html for more details.

LEDA[10] developed at Max Planck Institute für Informatik, provides a sizable collection of data types and algorithms in a form which allows them to be used by non-experts. It is implemented in an object-oriented style with a C++ class library, and emphasizes data structure, algorithm reuse and demonstration.

Shastra[1] developed at Purdue University produces several interoperable, collaborationware toolkits for the creation, manipulation and reconstruction of solid models from a variety of sources. These sources include but are not limited to CT/MRI scans, 3D digitized point data and Oceanographic cartography data. These tools provide for the rapid prototyping and analysis of solid objects prior to simulation or manufacture. The applications include solid modelers, algebraic geometry systems, spline surface toolkits and medical image reconstruction systems as well as others. Current work involves tele-collaborative visualization of scalar and vector data from problems in fluid flow, medical imaging and image analysis as well as the development of virtual environments that allow for the simulation and analysis of various physical "worlds". For details see http://www.cs.purdue.edu/research/shastra/shastra.html.

XYZ GeoBench[13] developed at ETH, Zurich, Switzerland, is a unified geometric programming environment, providing tools for creating, editing, and manipulating geometric objects, and demonstrating and animating geometric algorithms. It provides a user interface as well as a library and is implemented in an object-oriented style running on MacIntosh. See http://wwwjn.inf.ethz.ch /group/projects.html for more information. A similar work, called Workbench, was developed at Carleton University[6].

3 Future Work

Here we briefly give some possible directions for future work in the context of geometric algorithm visualization and software development. It is a *wish-list* of things to have.

Software Library Currently we need *more* geometric primitives at all levels. LEDA does provide basic data types for various fundamental data structures, and a collection of graph algorithms. These primitives include routines for computing intersection of convex polyhedra, convex hull computation, Delaunay triangulation in d-dimensions, $d \geq 2$, triangulation of a simple polygon, just to name a few of those fundamental geometric algorithms. It will be useful if a default visualization or animation routine for each of these operations is readily accessible as easily as a function invocation. Such a common software library is desperately needed.

Rapid Prototyping To facilitate development of software tools or primitive functions a programming environment that provides facilities for performing unit testing and module integration is desirable. Visualization can be exploited to help algorithm developers visually examine and verify the output of each module. Developing geometric software is not much different from an ordinary software development task. One should follow software design principles and incorporate visualization into the design cycle to grasp the essence of the behavior of the geometric algorithm being developed.

Debugging Debugging is always a time-consuming process in any software project. It is more so in debugging geometric software. Traditional source code debuggers are all text-based. A *visual debugger* that can print the content of a geometric object in graphic form will be of tremendous value. Current tools, such as GASP and GeoSheet, for example, can provide assistance in this regard. But much needs to be done in the area of *interactivity* to improve the usage of the tools. See below.

Interactivity The visualization tools available to date are mostly data-driven. User interaction with these tools is minimal in the sense that the user can only change the data file and re-execute the visualization routine to visualize the altered data. The tools that allow the user to directly manipulate input data on the display via point-and-click mechanism are lacking. For instance, tools that permit the user to point to an object (point or otherwise) and modify the position of the object or the information associated with the object are not available as yet. It would be very useful to have this capability so that the user can directly manipulate the objects and observe the effect of these changes on the algorithm as the program is being executed without going through recompilation process.

Distributed Environment As the hardware becomes more powerful and accessible, it is recommended that availability of heterogeneous machines be taken into account when designing geometric code and visualization software. Remote execution/access capabilities across the internet will alleviate the burden of installation at the local site of software tools developed at remote sites. The concept of collaborative visualization discussed in [1] may prove useful here.

References

1. Anupam, V. C. Bajaj, D. Schikore and M. Schikore, "Distributed and Collaborative Visualization," *Computer*, 27,7 July 1994, pp. 37-43.
2. Brown, M. H., "Zeus: A System for Algorithm Animation and Multi-view Editing," *Proc. IEEE Workshop on Visual Languages*, Oct. 1991, pp. 4-9.
3. Chazelle, B, *et al.*, "Application Challenges to Computational Geometry," Tech. Report TR-521-96, Princeton University, April 1996. Also accessible on the Web at http://www.cs.princeton.edu/~chazelle/taskforce/CGreport.ps.
4. de Rezende, P. J. and W. R. Jacometti, "GeoLab: An Environment for Development of Algorithms in Computational Geometry," *Proc. Int'l Computational Geometry Software Workshop*, Geometry Center, MN, Jan. 18-20, 1995.
5. Edelsbrunner, H. and E. P. Mücke, "Three-Dimensional Alpha Shapes," *ACM Trans. on Graphics*, 13,1 Jan. 1994, pp. 43-72.
6. Epstein, P., J. Kavanagh, A. Knight, J. May, T. Nguyen, and J.-R. Sack, "A Workbench for Computational Geometry," *Algorithmica*, April 1994, pp. 404-428.
7. Fabri, A. G. Giezeman, L. Kettner, S. Schiia and S. Schönherr, "The CGAL Kernel: A Basis for Geometric Computation," this proceedings.
8. Lee, D. T., S. M. Sheu and C. F. Shen, "GeoSheet, A Distributed Visualization Tool for Geometric Algorithms," Tech. Rep. Department of EECS, Northwestern University, Oct. 1994. *Int'l J. Computational Geometry & Applications*, to appear.
9. Phillips, M., S. Levy and T. Munzner, "Geomview: An Interactive Geometry Viewer," Notices American Mathematic Society, 40,8 Oct. 1993, pp. 985-988.
10. Näher, S., "LEDA – A Library of Efficient Data Types and Algorithms", Max-Planck-institut für informatik.
11. Roman, G.-C. and K. C. Cox, "A Taxonomy of Program Visualization Systems," *Computer*, 26,12 Dec. 1993, pp. 11-24.
12. Roman, G.-C., K. C. Cox, C. D. Wilcox, and J. Y. Plun, "Pavane: A System for Declarative Visualization of Concurrent Computations," *J. Visual Languages and Computing*, 32 June 1992, pp. 161-193.
13. Schorn, P., "An Object Oriented Workbench for Experimental Geometric Computation, " *Proc. 2nd Canadian Conference in Computational Geometry*, Ottawa, August 6-10, 1990, pp. 172-175.

14. Stasko, J. T. "Tango: A Framework and System for Algorithm Animation," *Computer*, 23,9 Sept. 1990, pp. 27-39.
15. Tal, A. and D. Dobkin, "Visualization of Geometric Algorithms," *IEEE Transactions on Visualization and Computer Graphics*, 1, 2, June 1995, pp. 194-204.

Position Paper for Panel Discussion

Kurt Mehlhorn

I discuss four issues related to the construction of software libraries: correctness, elegance and ease of use, efficiency, and extendability. All items will be discussed on the basis of our experiences with LEDA [7, 9] (and CGAL).

Correctness: Programs which are put in a library must be correct. In the geometric setting this implies that they must be able to handle degeneracies and they must not brake due to numerical errors.

By the time of the conference CGAL and LEDA will offer two kernels for computational geometry, one for computations in two-dimensional space [3] and one for computations in arbitrary dimensional space. Both kernels use exact rational arithmetic (based on integer arithmetic and homogeneous coordinates). I hope that the kernels will considerably simplify the development of correct geometric software.

Some algorithms based on the kernels will also be available, e.g., line segment intersection, boolean operations on polygons, convex hulls, Vornonoi diagrams, and Delaunay triangulations. The algorithms will work for all inputs. [1, 10].

The kernels deal almost exclusively with linear objects (points, lines, planes). Non-linear objects, e.g., circles, are only supported as long as rational arithmetic suffices. However, many simple geometric problems go beyond rational arithmetic. The Voronoi diagram of line segments is one such problem. It requires to compute with algebraic numbers. Burnikel et al [2] have recently shown that the geometric tests required by this algorithm can be evaluated with reasonable efficiency. We are currently adding an exact algorithm for Voronoi diagrams of line segments to LEDA.

A correct algorithm is no guarantee for a correct program. Program checking may be used to increase reliability [8].

Elegance and ease of use: Another must. Programs written in a library must look good. They should be fun to read. This requires that we identify the correct geometric primitives and that we find elegant ways to hide implementation details. LEDA has done a good job with respect to data structures and graph and network algorithms. The problem is harder in geometry for two reasons. Our objects are more complex, e.g., should there be a convex hull data type, and our objects are not yet filtered through two generations of text books.

Efficiency: That's where our field is strong and feels strongly about. The kernels emphasize correctness over speed. A number of recent papers, e.g., by Fortune, vanWijk, Clarkson, Boissonnat et al., have demonstrated that clever use of floating point arithmetic can yield very efficient and yet correct implementations of some geometric primitives. We [6] have used some of these ideas in our line segment intersection algorithm. The results are very encouraging and we have a version of the two-dimensional kernel that uses floating point filters. We do not know how to incorporate the ideas into our higher dimensional kernel.

Most of our potential users do not feel so strongly about asymptotic running times. Also, those which feel strongly about efficiency are more interested in the question "Can I solve my typical instances in a second?"

Extendability: LEDA is now widely used and a large number of projects have been based on it. However, it is very difficult for a person outside MPI to contribute to the system itself. It retrospect I would say that this was beneficial to the project. It helped to make the system consistent and elegant. I also believe that we reached a state where we should make it easier to write extensions of LEDA. We have taken two steps in this direction.

- Stefan Näher and I are writing a book on the system. We describe its architecture, its implementation, and its use [5].
- We developed tools that support the construction of LEDA-style manual pages and documentation [4].

References

1. Ch. Burnikel, K. Mehlhorn, and S. Schirra. On degeneracy in geometric computations. In *Proc. SODA 94*, pages 16–23, 1994.
2. Ch. Burnikel, K. Mehlhorn, and St. Schirra. How to compute the voronoi diagram of line segments: Theoretical and experimental results. In Springer-Verlag Berlin/New York, editor, *LNCS*, volume 855 of *Proceedings of ESA '94*, pages 227–239, 1994.
3. A. Fabri, G.-J. Giezeman, L. Kettner, S. Schirra, and S. Schönherr. The CGAL Kernel: A basis for geometric computation. *To appear at Workshop on Applied Computational Geometry (WACG96)*, 1996.
4. E. Haak, K. Mehlhorn, S. Näher, and Ch. Uhrig. Specification and documentation in the LEDA system. to appear.
5. K. Mehlhorn and S. Näher. The LEDA platform for combinatorial and geometric computing. to appear.
6. K. Mehlhorn and S. Näher. Algorithm design and software libraries: Recent developments in the LEDA project. In *Algorithms, Software, Architectures, Information Processing 92*, volume 1, pages 493–505. Elsevier Science Publishers B.V. North-Holland, Amsterdam, 1992.
7. K. Mehlhorn and S. Näher. LEDA: A platform for combinatorial and geometric computing. *Communications of the ACM*, 38(1):96–102, 1995.
8. K. Mehlhorn, S. Näher, T. Schilz, S. Schirra, M. Seel, R. Seidel, and Ch. Uhrig. Checking geometric programs or verification of geometric structures. To appear at the 12th Annual Symposium on Computational Geometry, 1996.
9. S. Näher and Ch. Uhrig. The LEDA User Manual (Version R 3.2). Technical Report MPI-I-1-002, Max-Planck-Institut für Informatik, 1995.
10. Michael Seel. Eine Implementierung abstrakter Voronoidiagramme. Master's thesis, Max-Planck-Institut für Informatik, 1994.

Designing the Computational Geometry Algorithms Library CGAL

Mark H. Overmars

Department of Computer Science, Utrecht University, P.O.Box 80.089, 3508 TB, Utrecht, the Netherlands. Email: markov@cs.ruu.nl.

1 Introduction

The field of computational geometry has now advanced to a state where it is possible to create robust, efficient, and reusable implementations of most geometric algorithms. At the same moment, in many potential application domains for computational geometry, like computer graphics, robotics, geographic information systems, and computer vision, people are more and more realizing that concepts and algorithms from computational geometry can be of importance to their work. This, for example, shows from the fact that in the last year a number of companies explicitly asked for knowledge of computational geometry when hiring new personel. This indicates that there is a need for a well-designed and well-implemented library of geometric routines. The question though is who should implement such a library. For a long time, most people within computational geometry considered this the task of the people who wanted to use the algorithms, and not a task for the field itself. In recent years people have come to understand that this is not the way it is going to work. Geometric algorithms are difficult to understand and implement and people in the application domains normally don't have the knowledge, nor the time, to do these implementations. Hence, the field of computational geometry itself should produce such a library.

This note descibes the goals of the CGAL-project, a (mainly) European initiative to create a software library, called the "Computational Geometry Algorithms Library", that is currently underway. I will in particular discuss what type of decisions have to be made when designing such a library and what choices we made in CGAL. For more details on the CGAL-library see the paper *The CGAL Kernel: A Basis for Geometric Computation*, also in these proceedings.

2 The CGAL project

The CGAL-project is a joint initiative of seven sites: Utrecht University (the Netherlands), ETH Zürich (Switzerland), Free University Berlin (Germany), INRIA Sophia-Antipolis (France), Max-Planck-Institute Saarbrücken (Germany), RISC Linz (Austria), and Tel Aviv University (Israel). The goal of the project is to design and implement a robust, easy to use, and efficient C++ software library of geometric algorithms. The library is developed in close collaboration with 10 companies in Europe. The project will be funded by the European Union within the ESPRIT IV LTR programme.

This library will consist of a number of different parts. The kernel will contain the basic geometric data types and (constant-time) operations on them. The basic library will be a collection of implementations of a large number of standard geometric algorithms (convex hull, Voronoi diagram, etc.). The support libraries will interface CGAL-programs to other packages (e.g. for visualization, to perform I/O, etc.). Finally, there will be application extension in which collections of geometric routines are provided that are useful for some specific application domain (like geographic information systems, computer graphics, robotics, and computer vision).

The currect status (june 1996) of the project is that a first version of the basic part of the kernel has been implemented. This implementation is being evaluated and tested by the partners in the project. In the fall of 1996 a first public version will be made available. Next we will start on building the basic library and the support libraries. A first public version of these parts is expected to be available in the spring of 1997.

For more up-to-date information about CGAL and access to releases (when they become available) I refer the reader to the CGAL world wide web page: http://www.cs.ruu.nl/CGAL.

3 The user

When designing a software library the first question to ask is: who is going to use it. Different groups of users require a different type of library. Here the word "user" should be interpreted as the person who writes software using the CGAL-library; not the person who uses this application software. Potential users of CGAL can be found in different places. First of all there are the researchers working in computational geometry itself that want to use the library to more easily implement and test their own algorithms. Second, there are other researchers that want to use geometric algorithms in implementations done for their research. And third, there are people working in companies that want to use CGAL in commercial applications. All these groups of users have rather different demands.

People in computational geometry have a good knowledge of the algorithms in the library. Hence, they don't need much background knowledge and probably don't have any problems with adapting the code to fit their needs when required. They are often not interested in issues like platform independence, documentation, etc. They do though want an open system that they can easily change. They want full control over the way the library works. Also, they might not want to spend too much time on understanding the language aspects of the library. Instead they want fast and easy access to the parts they need.

Other people in research are most likely less familiar with computational geometry. Here it is essential that the software is well documented and specified. They most likely don't want to change code in the library but prefer to use it as a black box. They also don't want to spend a lot of time on learning how to use the library. They want easy-to-use building blocks. Platform independence is not so much an issue.

For commercial application the situation is quite different. Software libraries should first of all be reliable and there should be a guarantee for availability in the future. Spending a reasonable amount of time in learning how to use the library is acceptable as long as the library is general enough to be useful in many applications. The ability to easily interface the library to existing code is crucial here. Platform independence is also often important.

It will be hard to satisfy all needs in one library. In particular there is some sort of a conflict in generality and extendability on one hand and ease of use on the other hand. Also the more easy it is to change the internals of the library, the harder it is to guarantee reliability. Finally, platform independence often has a price in the form of reduced functionality.

In CGAL we have choosen to create a product of professional standard. Hence, we will primarily focus our work on users in commercial applications. We believe that such a library will also be useful for the other groups of users as long as we provide some additional support for visualization, filters, etc. (see below). To guarantee that the library does indeed meet professional standards we will closely collaborate with 10 companies who will test early version for usefulness in their environment.

4 Robustness versus speed

A well-known problem in the implementation of geometric algorithms is the question of robustness versus speed. In some application (e.g. fast graphics animations) correctness of the algorithms is of minor importance and speed is of major importance. In other applications (e.g. dimensional metrology) speed is not important but robustness is essential. So in some application one would prefer fast floating point arithmetic while in other applications one needs exact real arithmetic.

It is very difficult to write a library that will satisfy all needs. In particular, putting algorithms in an integrated library will always introduce a performance penalty. The fastest speed can only be obtained using dedicated direct implementations. Still there are, I think, some basic criteria that should always be satisfied. The first is that the routines should never break down due to robustness problems. Secondly, the routines should deliver what is written in the specification. For example, you might specify that a convex hull always returns a convex polygon but that it might be at most a given distance ϵ away from the actual convex hull. Alternatively, you might specify that the convex hull is a polygon with points of the set as vertices but is might not actually be convex. Such specifications will help the application programmer to make sure that his or her part of the code never breaks down. It is for example important to know whether the convex hull algorithm always returns a convex polygon or not.

In CGAL we give the user a choice of arithmetic. This is done used parametrized data types. It is even possible to use different types of arithmetic in different parts of the program. In this way time-critical parts can be done in floating point while precision-critical parts can be done using exact arithmetic. Globally switching the arithmetic is also easy. In this way one can e.g. temporarily switch to exact arithmetic to test what caused a particular bug in the program, while the production version uses floating point for reasons of speed.

5 Visualization

As computational geometry deals with geometric objects, there is obviously often a demand to visualize these objects. Also, users will often want to provide input in a visual way and in e.g. debugging, visual feedback is essential. An important question is how to incorporate visualization in a software library. There are two fundamentally different approaches: incorporate it in the library itself or create separate modules for this.

Incorporating visualization as part of the library is the more natural concept in an object-oriented setting. Each object simply has methods for e.g. drawing it and has certain properties that determine its visible appearance. The user could draw a red cube by writing something like

```
cube.setcolor(RED);
cube.draw();
```

The problem with such an approach is that all functionality for drawing must be part of the library itself. As a result the library will become huge, objects use more storage, the functionality will be limited, and it is difficult to extend. In particular, when using the library in a commercial setting, the users normally want to interface their programs to all sorts of other packages they use. Most likely, the built-in routines will not be adequate for this.

The complete opposite would be not to provide visualization tools at all but to enable the users to get enough information about the objects to do the visualization themselves. For example, when users can obtain the coordinates of the cube they can easily write the drawing routines. The big disadvantage here is that, unless the user is very careful, it will become extremely difficult to port applications to other platforms where other visualization tools must be used. Also, it will enforce many users to do the same type of work.

The solution we have choosen in CGAL is not to provide functionality for visualization within the library but to provide a general mechanism to interface CGAL to visualization packages. For this the C++ stream is used. The idea is that there are different classes of streams for different visualization packages. Each of these classes though has a same basic functionality. Additional functionality can be added depending on the possibilities of the package. The above example for drawing the cube would now look like

```
Postscript.stream mystream(...);
mystream << RED;
mystream << cube;
```

Using something else than Postscript can simply be achieved by changing the first line only. A number of stream classes will be provided (separately form the basic library) but users can create their own classes. Streams can also allow for input, depending on the type of visualization package used. Actually, this concept will also be used for other things than visualization, like conversion.

We believe that this approach to visualization provides both uniformity and openness while it does not clutter the library with unnecessary functionality.

6 Library or gems

Rather than constructing a library of geometric algorithms it is also possible to provide geometric algorithms as a collection of separate programs (or modules), much along the line of the well-known series of graphical gems. In graphics this approach has turned out to be very successful. The advantage is that everybody can contribute in an easy way. These pieces of code can then either be used as stand-alone programs (or modules) to perform a specific task, or they can be modified by the users to their specific needs.

Even though it would be highly useful to have such a collection of geometric gems available, I see a number of problems with this approach. First of all, the code for geometric algorithms is often very complicated. Hence, in most cases, it will be almost impossible for users to change the programs. So they will mainly be used as stand-alone programs or modules. But this requires a very carefully designed interface between the different modules. Computational geometry uses a huge number of data types that often need to be extended for particular applications. The different modules are only useful if they can handle this in a consistent way. A major problem here is that the data types, and hence the interface between the modules, should be defined in a very open way. For example, in applications few people simply want to compute the Voronoi diagram of a set of points. These points represent something, so they have addition data associated to it. Once the Voronoi diagram is constructed, this data should in some way be associated to the Voronoi regions (e.g., through the points). So the Voronoi diagram module cannot simply return a planar subdivision.

To enable this type of use, modules should be carefully implemented which makes it difficult to do this in the free, distributed way as is normal for such libraries of gems. Careful interface rules must be enforced. Also careful rules on documentation and specification are required. It is doubtful whether people are indeed willing to adopt their implementations to these rules. This is why we think that, in particular in computational geometry, an integrated library approach is better. Of course, once the kernel and basic part of the library are ready, it is rather easy for people to create independent modules for various problems, that, as long as they use the data types and basic routines from the library, can easily be used by other people. These would then form a sort of gems, but based on a solid foundation.

A main disadvantage of an integrated library is that it is not very appealing for somebody who just needs something simple. For example, if in an application only at one point you need to compute a convex hull, you most likely don't want to learn a lot about CGAL, change all your data types accordingly, etc. To this end we plan to create a set of Unix filters for a number of basic geometric algorithms, using a simple common file structure. Moreover we will provide a simple library to read and write such files. This will enable these casual users to use the routines without the need to learn CGAL.

7 Conclusion

We believe that the way we have set-up the CGAL-project will have a big chance of producing a reliable, efficient, versatile, and relatively easy to use library of geometric algorithms. We hope that CGAL will eventually become widely used, both in academia and in industry. To make sure that the library fulfils the needs of many users, we do encourage everybody to comment on our ideas, such that, if required, we can adapt the design in an early stage. You can mail your comments to cgal@cs.ruu.nl or give them through the world wide web page http://www.cs.ruu.nl/CGAL.

The Computational Geometry Impact Task Force Report: An Executive Summary

Bernard Chazelle

Department of Computer Science, Princeton University, Princeton, NJ 08544, USA.
chazelle@cs.princeton.edu

Abstract. In June of 1994, I invited a group of researchers[1] to join in an effort to assess the impact (past, present, and future) of computational geometry in the practice of geometric computing. With rapid advances in computer hardware and visualization systems, geometric computing is creeping into virtually every corner of science and engineering, from design and manufacturing to astrophysics to molecular biology to fluid dynamics. Can computational geometry meet the algorithmic needs of practitioners? Should it look to applied areas for new sources of problems? Can CG live up to its potential and become a key player in the vast and diverse world of geometric computing? These are some of the questions addressed in the Computational Geometry Impact Task Force Report. The document was prepared by a group of computer scientists, engineers, and mathematicians with extensive experience in geometric computing. The report was intended as a wake-up call rather than an agenda setter, meant to engage a community-wide discussion on the future of computational geometry. What follows is a brief executive summary of the report. The full report is accessible at the URL, http://www.cs.princeton.edu/~chazelle/taskforce/CGreport.ps.Z. It is also available as Technical Report TR-521-96, Princeton University, April 1996.

---

[1] The Computational Geometry Impact Task Force: Nina Amenta (Xerox PARC), Tetsuo Asano (Osaka Electro-Comm. U.), Gill Barequet (Tel Aviv U.), Marshall Bern (Xerox PARC), Jean-Daniel Boissonnat (INRIA), John Canny (U.C. Berkeley), Bernard Chazelle (**Chair**, Princeton U.), Ken Clarkson (AT&T Bell Laboratories), David Dobkin (Princeton U.), Bruce Donald (Cornell U.), Scot Drysdale (Dartmouth U.), Herbert Edelsbrunner (U. Illinois at Urbana-Champaign), David Eppstein (U.C. Irvine), A. Robin Forrest (U. East Anglia), Steve Fortune (AT&T Bell Laboratories), Ken Goldberg (U.C. Berkeley), Michael Goodrich (Johns Hopkins U.), Leonidas J. Guibas (Stanford U.), Pat Hanrahan (Stanford U.), Chris M. Hoffmann (Purdue U.), Dan Huttenlocher (Cornell U.), Hiroshi Imai (U. Tokyo), David Kirkpatrick (UBC), D.T. Lee (Northwestern U.), Kurt Mehlhorn (Max Planck Inst.), Victor Milenkovic (U. Miami), Joe Mitchell (SUNY at Stony Brook), Mark Overmars (U. Utrecht), Richard Pollack (Courant Institute, NYU), Raimund Seidel (U. Saarbrücken), Micha Sharir (Tel Aviv U. and NYU), Jack Snoeyink (UBC), Godfried Toussaint (McGill U.), Seth Teller (MIT), Herb Voelcker (Cornell U.), Emo Welzl (ETH Zürich), and Chee Yap (Courant Institute, NYU).

1 Preamble

The fraction of computing falling under the loosely defined rubric of "geometric computation" has been on the rise and is likely to become dominant in the next decade. Computer graphics, manufacturing, scientific visualization, computer vision, astrophysics, molecular biology, and fluid mechanics are just a few in a crowd of avid users of geometric computing. Where does computational geometry fit into all this?

Twenty-odd years ago, the nascent field of computational geometry set sail on a mission to build general tools — analytical and computational — to satisfy the algorithmic needs of geometric computing. The intention was to create a body of knowledge to which computer programmers could turn for help when wrestling with geometric problems. The vision was that of a two-way pipeline: applications areas feeding CG with important practical problems, and CG in turn providing answers in the way of algorithmic tools and mathematically sound analyses.

What is the state of the "pipeline" today? Before answering this question, one may want to step back a little. Any new field needs to build itself a home and a home needs foundations. On that score one may strike a celebratory note. Indeed, computational geometry has met with considerable success in resolving the asymptotic complexity of basic geometric problems (e.g., convex hulls, Voronoi diagrams, triangulations, low-dimensional optimization). Most of the classical problems posed in the first decade of the field have been solved. Many of them entailed building entirely new sub-areas in the field of algorithm design. For example, the twin use of randomization and derandomization has been the main force behind some truly remarkable success stories (e.g., optimal algorithms for convex hull and Voronoi diagrams, linear-time solutions of LP-type problems). Multidimensional searching, triangulation, geometric sampling, the theory of Davenport-Schinzel sequences, and zone theorems are some of the crowning intellectual achievements in the field of CG.

Computational geometry has been remarkably dynamic, productive and creative. In fact, in the area of algorithms and data structures, CG is arguably where most of the action has been in the last decade. Building theoretical foundations was undoubtedly the proper course to chart at the outset: CG can look back to its accomplishments with legitimate pride.

In the midst of its success, however, the field is standing at a crossroads. There are two options: CG can use its successes as justification for keeping the pursuit of theoretical investigations as the centerpiece of its agenda. Or it can move towards building an effective pipeline with geometric computing. In the first case, CG might simply fall out of the economic loop altogether: it might even shrink to the status of a recreational activity. In the second case, CG might grow to become as indispensable to geometric computing as, say, civil engineering is to bridge-building. The choice is for the CG community to make.

Why such alarm? The sobering reality is that, notwithstanding a few exceptions, the pursuit of practical solutions that specifically address the users' needs has lagged behind. At the same time, the computational geometry community has been largely unsuccessful in reaching out and communicating its more practical discoveries to potential users. Technology transfer has been slow and sporadic. To put it bluntly, the much-vaunted pipeline looks more like a pipe dream.

It need not be so. An effective pipeline is probably a feasible project: one certainly worth undertaking given the stakes. Among the problems to overcome first are the "obvious" ones: CG'ers must start thinking not only in terms of asymptotic complexity but also in terms of code robustness, precision, CPU times, standards, benchmarks, distribution, etc. Perhaps less obvious, though just as important, are communication problems of a more cultural nature: For example, why is it that in the graphics and computer vision literatures, rotating an object is considered (rightly so) a linear algebra problem, but computing a visibility graph or matching a geometric pattern is rarely considered a computational geometry problem? There are many computational-geometric questions in vision, graphics and robotics that are not recognized within those fields to belong in the province of CG'ers.[2] For this CG deserves most of the blame. It has failed to address the problems that people in practice want to see solved. In the absence of a pipeline, CG can never hope to be the place practitioners turn to for help. The sobering fact is that, realistically, the pipeline is mostly for CG'ers to build.

Does geometric computing need CG? It is too easy and tempting for CG'ers to convince themselves that it is the case. We try not to succumb to this temptation here. We simply state our belief that computational geometry is the *only* arena dedicated solely to algorithmic pursuits in geometric computing: discrete geometers want to understand countable geometric structures; combinatorial geometers want to count or approximate or enumerate them; practitioners want to produce working code for solving specific geometric problems. Only computational geometers make the algorithmic understanding of geometric problems their central preoccupation. We believe such pursuits can be the key to breaking current and future computational bottlenecks in many areas of engineering. This document gives abundant evidence to support this view.

If so, then why has CG broken so few "computational bottlenecks" that practitioners care about? Perhaps CG has been reluctant to cast itself in the demeaning role of "algorithm caterer" at the service of practitioners. This narrow view of CG is both false and dangerous. CG's link to the real world is essential to its intellectual vitality. The current feebleness of that link has been harmful to the field as a whole. Not only has it squelched the impact of CG in the engineering community, but research opportunities have been missed, scientific challenges have been overlooked. The intellectual landscape of CG has much to

---

[2] Defined loosely as the community gathered around such conferences as ACM SOCG, CCCG, and journals such as D&CG, IJCG&A, CGT&A, or more broadly, the geometric components of Algorithmica, J. Algorithms, SIAM J. Computing, JACM, etc. We recognize that many people not attached to that community do bona-fide computational-geometric work nevertheless. The focus of this report, however, is on the community that makes the design and analysis of geometric algorithms its primary occupation.

gain from close interaction with the real-world practice of geometric computing.

It is not our intent to pit theory and practice against each other, and decree the relative worthiness of each. We also want to make it absolutely clear that we are firm believers in the crucial importance of theoretical work. If anything we believe that the existence of an effective pipeline would further motivate all sorts of new theoretical investigations which the current insularity of the field has hindered. Our view is that a mature field of computational geometry should be home to a diversity of interests, some theoretical and some practical, most of them feeding on and contributing to the others.

It is particularly important to create a platform linking theoretical research at one end to production-quality, usable software at the other end. Such a platform exists in combinatorial optimization, where some of the most impressive theoretical advances (such as interior-point methods) have had tremendous impact in practice. CG is nowhere near this happy state of affairs.

To get there will require a concerted effort. The prevailing winds are favorable, however. A combination of factors, from the intellectual maturing of CG to more prosaic forces such as the job market and funding agencies, are causing a subtle but profound reorientation of the field. Instead of letting these factors alone dictate this reorientation, this document attempts to make the CG community itself the main agent of reform.

Computational geometers are blessed with a field whose scope dwarfs most other areas of computer science. Cryptography, graph algorithms, optimization, computational biology, etc., are all critically important, but they are relatively well-focused areas. By contrast, computational geometry spreads its wings over the entire computing spectrum. This breadth is an asset. But it is also a challenge. Indeed, the field is so vast that it is fragmented and has trouble recognizing itself under one roof. Computational geometry is diverse and so is its community. Within this diversity, however, we must strive to build an identity and unity of purpose. This is the main objective of our effort.

This is not meant to imply that all worthy computational-geometric research is to be done under the aegis of the CG community. Researchers in graphics, modeling, manufacturing, biology, etc, have their own traditions of fine geometric work: there is no need for this to change. What must be remedied is the lack of communication between practitioners and theoreticians of geometric computing.

It is imperative to make CG research more responsive to the needs of users. But, again let us restate our belief that not all computational geometers need to work in directly applicable research. Long-term structural explorations are also needed. Our point is not to emphasize practice to the detriment of theory. On the contrary, our chief goal is to broaden —not simply to shift—the band which CG occupies in the spectrum of computer science: Instead of a collection of small houses scattered across town and oblivious to one another, we envision CG as a large edifice with a theoretical wing and a practical wing and many aisles (the "pipeline") connecting them. Our purpose is not to bring down one of the wings, but to consolidate the edifice.

2 Recommendations

It will take more than a change of heart to reform computational geometry. Structural changes are necessary. Happily some of them are already underway. We identify four broad categories: (1) Production and distribution of usable (and useful) geometric codes; (2) Joint forums between CG and applied areas; (3) Experimental research; (4) Reform of the reward structure in CG.

Remark: Educational matters were left out from our discussion. Teaching and research in CG at the undergraduate and graduate levels are critical issues that must be addressed. But the great differences among national education systems precluded a discussion that would have been of much meaning to the international CG community at large.

1. PRODUCTION AND DISSEMINATION OF GEOMETRIC CODE: Journals are the time-honored repository of scientific knowledge. But what about software? Journals are woefully inadequate. An on-line library of geometric code available through the Internet would be a useful starting point. Many questions need to be ironed out. Should the library conform to rigid formats and should contributions be refereed (ie, pass a number of standardized tests) before being archived? Or should it follow the freewheeling, open-door policy of most archives on the Web? What programming languages should be used? What documentation and maintenance service should be expected? Should data type conventions be enforced?

   Applied areas such as graphics and geometric modeling have produced vast amounts of geometric code over the years, some of it publicly available. Sometimes, differences in programming languages and standards might make integration difficult, but this need not be always the case. At the very least, how to tap into such resources effectively should be addressed. Our sense is that many of these issues cannot be settled once and for all before trying out various alternatives on a smaller scale. There are models to learn from in optimization, computer algebra, and numerical analysis. But it is unclear that any of them can be replicated verbatim to fit the needs of CG. For example, the uniquely complex data typing in CG places formidable hurdles in the way to correctness, robustness and portability. Only preliminary efforts have been attempted in CG so far, and to plan *the* perfect system in a vacuum is perhaps utopian. People should feel encouraged to try out their ideas and time will tell what works and what does not.

2. INTERDISCIPLINARY FORUMS: CG'ers should attend conferences in applied areas and practitioners should be invited to address the main CG conferences. This is already happening. Clearly, this is not enough. To provide fresh grist for CG's mill, means must be found so that CG'ers are given opportunities to hear about computational-geometric bottlenecks in applications areas. For example, one could organize joint conferences between CG and specialized topics. Special issues of non-CG journals could be devoted to such meetings. At the same time, the potential users of CG'ers' output must be kept informed of new developments. A series of "CG gems" books pat-

terned after the computer graphics gems might be useful. Or at the least the graphics gems books, which already include significant amounts of geometry, should be given greater attention (and perhaps receive more contributions) by CG'ers. The current CG newsletters and mailing lists might spawn special wide-distribution issues (announcing newly released software or recent developments of interest to practitioners).

3. EXPERIMENTATION: To feed the CG pipeline with usable results, experimental work must be integrated within mainstream CG research. Actually, it is one of the great assets of CG that it lends itself so naturally to experimental work. Regrettably, that asset has been grossly neglected. The standard programming cycle (design, code, debug, test and benchmark) is painfully slow, so the number of iterations tends to be small. The field of optimization has shown that practical algorithmic innovations tend to require repeated iterations through the programming cycle. Because this does not happen often enough in computational geometry, algorithmic innovations tend to be mostly of a theoretical nature. The unusual slowness of the implementation cycle is due to the complexity of geometric data types and the lack of adequate software tools (code libraries, visual debuggers, dataset catalogs, etc.).

Quality experimental research (as practiced in, say, biology or physics) must satisfy highly demanding criteria. Any geometric algorithm that is claimed to be the "method of choice" should be not only implemented and tested, but benchmarked against its competitors. Claims of strengths should be backed up by credible evidence and weaknesses should be identified. None of this is possible as long as every piece of code must be written from scratch, as long as every test input must be produced by hand, and as long as every debugging, visualizing, and measuring tool must be hand-crafted. There is a need for a large-scale effort in building software tools for computational-geometric experimentation. Standards should be set by which to judge the quality of experimental work.

To allow for benchmarking, representative data sets should be collected and archived. Input data should enable effective robustness and efficiency testing. Raw data should be included as well as highly structured or datatyped data.

One might ask: experimental computational geometry in applied areas has been alive and well for years. Why do we need to create anew something which already exists? The answer is that in many cases experimental work has been so closely tied to applications that more general pursuits might have been overlooked. An analogy would be that although the oil, food, and pharmaceutical industries each pursue their own brand of experimental chemical engineering tailored to their needs, few chemical engineers would deny the value of unfettered experimention. The same is true in computational geometry. Unfettered geometric experimentation will complement (not supplant) ongoing experimental work in applied areas.

4. REWARD STRUCTURE. Conferences, workshops and journals should be receptive (as some already are) to experimental work and software building.

One possible suggestion is to run an annual workshop devoted solely to experimentation in CG, featuring geometric results as well as software tools. Researchers in relevant applied fields with a tradition of experimentation (like drug design) could be invited to share their experience with CG. For an experimental culture to take hold, it is essential that quality experimental work should be rewarded through the standard channels (journals, conferences, hiring, promotion, grant awards, etc.) For this to happen, experimental research must be judged and evaluated according to recognized standards. Of course, this is a bit of a chicken-and-egg question, and a certain amount of bootstrapping might be necessary. It probably means that quality standards should be made flexible enough at first to reflect the relative immaturity of CG experimentation. Within a few years, however, firm standards should be put into place.

Together with an experimental culture, a software systems culture must be encouraged to grow within CG. Building novel geometric software that transforms the practice of geometric computing should be considered on par with proving a theorem that changes the mathematical landscape of the field. The reward structure must be adjusted, and CG must learn how to judge non-theoretical research. This is not to say that one should systematically reward any attempt at writing code. On the contrary, yardsticks for distinguishing bad from good code should be introduced, and the standards should be just as rigorous as they are for evaluating mathematical work. Promotion and hiring decisions should reflect these cultural changes. Again, none of these changes can be decreed. A new way of thinking must first take hold.

Finally, CG needs not only to open up to experimental and software-building work but also to rethink its approach towards theoretical research. Several fundamental theoretical questions remain open in geometric optimization, combinatorial geometry, geometric primitives, geometric searching, etc. But the list of open problems whose centrality is so compelling as to draw a consensus –even within the CG community– is not nearly as long as one might think. We believe that many of the more interesting theoretical questions have not been formulated yet. These questions will surface as firmer bridges between CG and applied areas are created. For example, fundamental questions in computational topology might arise from geometric work in biology or astrophysics. Many of the classical computational-geometric concepts, such as convex hulls and Voronoi diagrams, arose from exposure to the natural sciences. Because CG is not a "foundational" science but a part of applied mathematics, it must draw its main inspiration from the physical world that it tries to model.

While CG must look outside for new frontiers to conquer, it must also become more critical of its current theoretical research. Problems whose only merit is to be open should probably be left as such. There are so many more open problems than what CG can ever hope to solve that it should focus on problems of identifiable importance, be they theoretical or practical.

Geometric Manipulation of Flexible Ligands*

Paul W. Finn[1], Dan Halperin[2]**, Lydia E. Kavraki[2], Jean-Claude Latombe[2],
Rajeev Motwani[2], Christian Shelton[2], Suresh Venkatasubramanian[2]

[1] Pfizer Central Research, Sandwich, Kent CT13 9NJ, UK
[2] Department of Computer Science, Stanford University, Stanford, CA 94305, USA

Abstract. In recent years an effort has been made to supplement traditional methods for drug discovery by computer-assisted "structure-based design." The structure-based approach involves (among other issues) reasoning about the geometry of drug molecules (or *ligands*) and about the different spatial conformations that these molecules can attain. This is a preliminary report on a set of tools that we are devising to assist the chemist in the drug design process. We describe our work on the following three topics: (i) geometric data structures for representing and manipulating molecules; (ii) conformational analysis—searching for low-energy conformations; and (iii) pharmacophore identification—searching for common features among different ligands that exhibit similar activity.

1   Introduction

Most existing pharmaceutical drugs were found either by chance observation or by screening a large number of natural and synthetic substances [7]. In recent years there has been a growing tendency to supplement the traditional methods of drug discovery by *structure-based design*. The structure-based approach builds on the improved understanding of the molecular interaction underlying diseases, and attempts to predict the structure of a potentially active compound. The prediction can then be used either to synthetically construct such a compound, or to narrow down the screening process of existing substances. Refer to [3] for a comprehensive survey and bibliography on the subject.

The fundamental assumption of structure-based drug design is that at the molecular level, the key event leading to the desired effect of the drug is the recognition and binding of a small molecule (the *ligand*) to a specific site on a target macromolecule (the *receptor*) [3]. It is further assumed that at the

* Work on this paper has been supported by a grant from Pfizer Central Research. Work on this paper by Rajeev Motwani has been supported by an Alfred P. Sloan Research Fellowship, an IBM Faculty Partnership Award, an ARO MURI Grant DAAH04-96-1-0007, and NSF Young Investigator Award CCR-9357849, with matching funds from IBM, Schlumberger Foundation, Shell Foundation, and Xerox Corporation. Work reported in Section 2.2 has been supported by ARO MURI Grant DAAH04-96-1-0007.
** Author's current address: Department of Computer Science, Tel Aviv University, Tel Aviv 69978, ISRAEL

binding site, the ligand must present steric and electrostatic complementarity to the receiving pocket.

Depending on whether the structure of the receptor is known or not, two types of structure-based approaches can be considered. If the structure of the binding site is known, the process then centers on finding (or devising) a ligand that will complement the binding site. Our work focuses on the case when the structure of the receptor is not known. In this case, a possible scheme is to start with a set of compounds whose structure is known, and which have been observed to exhibit some level of activity with the target receptor molecule. The goal then is to extract the common features of the active compounds. These common features constitute a *pharmacophore*. With a pharmacophore at hand, the chemist can look for other compounds having similar features, but which may be more potent than any of the given compounds, or have other desirable properties such as non-toxicity.

A major source of difficulty in structure-based drug design is the flexibility of molecules to attain various *conformations*, namely different spatial configurations of their atoms. Properly handling flexible ligands has been identified as a major challenge in this field [22].

We have chosen to concentrate on the following topics where algorithmic tools are needed to support the design process: (i) data structures for representing molecule geometry and molecular surfaces; (ii) conformational analysis — searching for low-energy conformations; and, (iii) pharmacophore identification. The first two components can be viewed as support tools for the pharmacophore identification part. Although our current effort focuses on tools to support the drug design process when the receptor structure is not known, some of the tools that we are developing can also be used as building blocks to support the design when the receptor structure is known.

There is an abundance of software tools for drug design [3, 4]. Many algorithms with a geometric flavor have been proposed and implemented in this domain. We believe that our work is innovative in the following aspects. In our study of data structures for representing molecule geometry, we aim to develop techniques that are efficient under conformation change, viz., techniques that will allow for efficient dynamic update of the structure as the molecule conformation changes. We have devised and analyzed several models for efficient dynamic maintenance of such structures, as discussed in Section 2. The goal of conformational search, which is discussed in Section 3, is to produce low-energy conformations of ligand molecules. These conformations constitute the input to the pharmacophore identification procedure. Conformational search is computationally expensive and may take hours on high speed workstations [21]. We aim for speed and efficiency of calculation, and are willing to find low-energy conformations that are not necessarily energy minima. Finally, in the pharmacophore identification component, described in Section 4, we consider a set of active molecules, each of which may be present in many low-energy conformations, and use inter-atom distances and molecular surface information to identify common structural elements of these molecules. Our search for pharmacophores is guided by user-specified size and accuracy requirements, and our software allows for an interactive refinement of the solutions obtained.

2 Geometric Data Structures

A prevailing approach to modeling the geometry of static molecules is to represent each atom as a ball of fixed radius in a fixed placement relative to the other atoms [25]. The radius assigned to each atom depends on the type of the atom. There are various sets of recommended values for atom radii, and a prevailing set is known as the *van der Waals* radii. In spite of its limitations, this model, which we will refer to as the *hard sphere model* of a molecule, is widely used.

Various techniques in drug design use *molecular surfaces*. One type of molecular surface is simply the outer boundary of the union of the balls (or spheres) in the hard sphere model above. This type is often referred to as the van der Waals surface. There are two closely related types of surfaces: the solvent accessible surface [23] and Richard's smooth analytical surface [28]. See also [9, 10, 11] and the survey by Mezey [25] for an extensive discussion on molecular surfaces.

A basic question in the geometric manipulation of molecules is the following: Given a hard sphere model of a molecule M and a query atom q, report the atoms of M intersected by q. It has been shown [16] that for a molecule with n atoms, an efficient data structure requiring $O(n)$ storage space can be constructed such that queries of the above type can be answered in $O(1)$ time each, after $O(n)$ expected preprocessing time. This data structure was used [16] to obtain efficient algorithms for constructing molecular surfaces and for computing the visibility map of molecules.

In the next subsection we describe an implementation of the above algorithm for computing the molecular surface. In Section 2.2 we extend the intersection data structure to the dynamic case.

2.1 Molecular Surfaces — Construction and Visualization

We construct an analytic representation of van der Waals molecular surfaces (the same procedure applies to solvent accessible surfaces as well). The representation consists of the patches that each atom sphere contributes to the outer surface, as well as the adjacency relations between patches on neighboring atoms that share a common arc. The basic procedure in our implementation of the computation of molecular surfaces is the construction of the subdivision (or *arrangement*) on each atom sphere s induced by the circles of intersection of other atom spheres with s. The arrangement is constructed incrementally, by adding one circle at a time, and maintaining the *trapezoidal decomposition*[3] of the current arrangement. See Figure 1 for an illustration.

---

[3] The trapezoidal decomposition is a refinement of the arrangement of the circles by adding certain arcs of great circles through the poles. For more details on trapezoidal decomposition, see e.g [26]. In our implementation, we extend such arcs only from points of tangency of original circles with great circles through the poles.

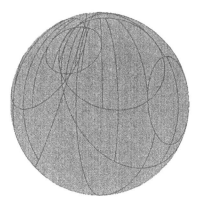

Fig. 1. Partial trapezoidal decomposition of an arrangement of small circles on an atom sphere.

We represent the arrangements of the circles using a quad-edge structure [13]. Once we obtain the arrangements for all atom spheres, we go on to construct the (van der Waals, or solvent accessible) molecular surface. This is achieved by adjusting the individual atom arrangements so as to retain only the faces in those arrangements that contribute to the outer surface, and by updating the edge information of edges that now border faces belonging to two different atom spheres (see [16] for details).

The arrangements on the spheres are used to compute the molecular surface area and the (possibly null) area contributed by each atom to the outer surface, as well as for graphic display of the intersection pattern of each atom with its neighbors (Figure 1). The information about area contribution of individual atoms is used in the pharmacophore identification module described below. Besides computing the outer surface, the program computes internal boundary components which bound *voids* (see, e.g., [11]) and outputs their surface area as well.

The implementation provides extensive facilities for interaction. It displays the molecule and the individual atom arrangements graphically, and also allows for re-coloring of the atoms according to several parameters such as the level of contribution to the outer surface. The edges of the arrangement on each atom sphere are colored to distinguish between the contribution of the atom to the outer molecular surface and to each void (if any). The program also provides various statistics on the data structures it uses and allows for experimenting with and fine-tuning certain parameters that control these structures.

2.2 Dynamic Maintenance

We can model the flexibility of a molecule to attain different conformations as if the atoms were rigid links of a robot linkage, and the bonds between some of the atoms are rotational (or rotatable) joints (see Section 3 for more details on this simplified model). We describe these kinematic constraints by a graph where each node corresponds to an atom sphere, and an edge between nodes describes a constraint. An edge can be either rigid — when there is a fixed relative displacement between the atoms that it connects, or rotatable — when there is a degree of freedom of rotation around a fixed line between the two atoms.

Suppose that we are given a sequence of update requests that are aimed at changing the conformation of the molecule. This is done by giving a sequence of joint angles to which we need to update the rotatable bonds. The sequence of updates is interleaved with intersection queries of the form: does a given query sphere intersect any of the atom spheres in the molecule at its current conformation? We require a strategy for maintaining a data structure which processes a sequence of updates and queries in optimal time, where the processing algorithm has the freedom to break or merge substructures.

In [15] we study several models of dynamic maintenance of such kinematic structures, and devise maintenance algorithms for them. We give a worst-case optimal strategy for the case where the molecule graph is a tree. The key idea behind this is that of a *balanced decomposition* of the tree into subtrees such that the number of subtrees (which corresponds to the cost of a query) is roughly the same as the size of each subtree (which corresponds to the worst case cost of an arbitrary update). We describe an efficient algorithm for constructing such a balanced decomposition. We also show that obtaining the optimal solution for a given sequence of updates and queries, even when the molecule graph is a path, is an NP-hard problem and we present approximation algorithms for this case.

3 Conformational Search

Searching the conformational space of small ligands is an important operation in the process of pharmaceutical drug design [22]. Given a function that computes the energy of the molecule, the goal of the search is to produce low-energy conformations that are geometrically distinct. These conformations can provide the input to a pharmacophore generation procedure, or can be used to screen large databases of protein molecules for possible docking receptors.

When conformational search is conducted, the prevailing practice is to consider only the torsional degrees of freedom of the molecule (see Figure 2 for an example). Other degrees of freedom, such as bond lengths between atoms or bond angles between consecutive bonds, are often ignored since their variation does not drastically change the molecular conformation. Another widely used approximation is to consider that the molecule is in vacuum. In this case, the energy of the molecule can be computed by empirical force-fields which consider only intramolecular quantities. The definition of such fields has been the result of intensive research [21]. Typically they have terms that involve pairs, triplets, and quadruples of bonded atoms, as well as pairs of distant non-bonded atoms. In reality, however, the conformation of a molecule is influenced both by intramolecular and intermolecular forces. Our techniques require as input a pro-

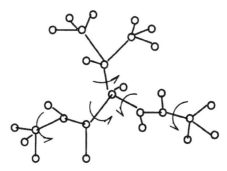

Fig. 2. The molecule of valine with its torsional degrees of freedom. A ball indicates an atom center, and a line segment indicates a bond between two atoms.

cedure that computes the energy of a molecule and will work if this procedure models the molecule in solution, or takes into account an external force field.

Conformational analysis is a very hard search problem when the molecule under consideration has more than 5 degrees of freedom. For a comprehensive survey of previous methods see [8, 21, 30]. During search, emphasis is placed on generating geometrically different conformations of a molecule within a user-defined energy interval. The underlying assumption is that one of these conformations will be adopted inside a receptor cavity. Selective generation of conformations with certain properties may also be desired. For example, if a pharmacophore is known, one may want to find low-energy conformations that retain the features of the pharmacophore at their relative positions. We describe below a method for generating low-energy conformations of a molecule, and our preliminary efforts to efficiently organize the resulting conformations in clusters and provide input for our pharmacophore generation procedure.

3.1 General Framework

Inspired by our success with probabilistic techniques for robot motion planning [17], we have implemented a search procedure which randomly samples the conformational space of small molecules. The energy of the molecules subjected to conformational search is computed by the Tripos Force Field [32]. Our method is divided into three steps: generation of random conformations, minimization of these conformations, and grouping or clustering of the minimized conformations.

Generation of conformations. During this step, a large number of conformations, frequently tens of thousands, are generated at random over the conformational space of the molecule. In contrast with previous search methods that discretize each torsional angle, we obtain a random conformation by selecting each torsional degree of freedom uniformly from its allowed range. However, if information is available for the preferred values of a particular torsion angle (such information has been collected from crystallographic databases [20]), then it is easy to select the value of this torsion angle according to a distribution that reflects the above information. The resulting structure is stored only if it avoids self-collisions which may result from intersections of the *hard spheres* of non-bonded atoms.

Minimization. An efficient minimizer [5, 29] is used to obtain conformations that are at local energy minima. It is important to note that minimization is the most time-consuming step during conformational search, and that its efficiency is crucial for the performance of the search. A large part of the running time is spent towards the end of the procedure, when the process tries to meet certain user-defined stopping criteria. We relax these criteria to achieve significant reductions in the running time at the expense of accepting conformations that are not at local energy minima. However, experiments show that conformations very close to a local minimum usually have similar geometries. These conformations are later grouped by our implementation in clusters as explained in the next paragraph. For the purposes of most applications (e.g., pharmacophore generation) only one conformation per cluster is retained. Hence, no relevant information is likely to be lost by avoiding the full minimization of conformations that belong to the same cluster. In the framework of conformation generation and minimization, the work described in Section 2.2 may prove beneficial in reducing the time spent on collision checking and on the calculation of various energy function terms.

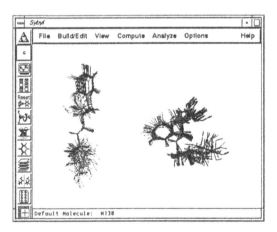

Fig. 3. Two clusters of an actual drug molecule.

Distinct geometric conformations. Interpreting a set of a few thousands of conformations is challenging and can benefit from extensive research in the area of clustering [18]. Currently, our system partitions conformations into clusters by using an easily-computable measure of distance between conformations. In particular, three distinct atoms a_1, a_2, a_3 are first selected in the molecule. The distance of two conformations is defined as the sum of the Euclidean distances of their corresponding atoms, after identifying atom a_1, direction a_1a_2, and plane $a_1a_2a_3$ in the two conformations. Clustering is performed by placing a given conformation in an existing cluster if its distance from the "center" of that cluster is less than a predefined value. If no such cluster is found, a new cluster is initiated. The center of a cluster can be the conformation with the lowest energy in the cluster. Despite its simplicity, this procedure seems to perform well in practice when a few thousands of conformations are involved. Examples of clusters for a molecule with 11 degrees of freedom are shown in Figure 3. We plan to investigate further the relationship between cluster generation and the extent to which minimization should be performed. Additionally, randomization and hashing techniques may reduce the complexity of clustering and permit its efficient use with very large numbers of conformations.

4 Pharmacophore Identification

In the literature, pharmacophore identification techniques model a molecule either as a graph (where atoms are vertices and bonds are edges) [31], or as a set of points in 3-dimensional space [6]. The second approach is often justified by arguing that pharmacophores tend to be non-local — bond information seems to have less of an influence on ligand-receptor binding than other properties such as hydrophobicity, hydrogen bonding, and electrostatic interactions (e.g., see [12]). Therefore, we model a molecule as a set of *labeled* points in three dimensions, where each point represents the center of an atom. In fact we expand each point into a small sphere of radius ϵ centered at that point, where ϵ represents uncertainty in locating atoms. Labeling allows us to generalize the notion of compatibility between atoms, so that we can take into consideration other structural/chemical properties of the atoms without having to change our algorithms. One type of labeling that we use is the surface area contributed by an atom to the outer molecular surface (see Section 2.1).

Since each molecule may exist in one of many low-energy conformations, the representation of a molecule consists of a collection of one or more point sets, each corresponding to one such conformation. The problem now is to determine one or more point sets that are congruent to a subset of some conformation in most of the molecules, where the only transformations permitted are translations and rotations. Note that allowing molecules to have multiple conformations makes the problem much more difficult, because now for any pair of molecules, we may potentially have to compare each possible pair of conformations in these molecules to determine even one pharmacophore.

It is known that if we restrict each molecule to one conformation, finding the largest such subset is NP-hard; from the point of approximate solutions, it is known to be hard to approximate [1] and only weak positive results are known [19]. There exist polynomial-time algorithms for the problem if we consider only two point sets [2], and these could be used to build algorithms over a larger number of point sets; however, these solutions are inefficient. In the next subsection, we describe an implementation based on a fast heuristic to determine pharmacophores between two point sets. Subsequently, we outline a scheme whereby such pharmacophores could be used to reconstruct a solution for many point sets efficiently.

4.1 The Two-Set Problem

Our implementation uses the following observation: if we can identify three pairs of atoms (one from each of the two conformations) that belong to a pharmacophore, then we can simply compute a transformation from the first triplet to the second. This transformation can be used to transform the first conformation, and a nearest-neighbor search will then yield the rest of the pharmacophore. We are not interested in very small pharmacophores, so we assume that any common substructure is not too small: specifically, for some constant $\alpha \leq 1$, there exists a common substructure P in conformations C_1, C_2 s.t.

$$|P| \geq \alpha \min(|C_1|, |C_2|).$$

This is a reasonable assumption because conformation sizes are typically of the order of 30-50 atoms, and non-trivial pharmacophores are of the order of 5-15 atoms. Therefore, α can vary between 0.1 and 0.3.

Now, we can randomly sample triplets of points from the smaller conformation. If we store distances between pairs of conformation atoms in a hash table, we can probe this table with the inter-point distances of the given triplet to determine triplets in the second conformation congruent to it. Then, as noted above, computing the transformation from one triplet to the other (if one exists) will yield a common substructure. We do this procedure repeatedly, until we obtain a substructure that satisfies the minimum size requirement.

It is easy to see that the probability of a random triplet belonging to a pharmacophore is at least α^3. Therefore, $\tilde{O}(\frac{1}{\alpha^3})$ iterations will yield the pharmacophore with high probability. In practice, we run the program in Monte Carlo mode – if a large invariant is not detected within these number of iterations, the program halts, reporting failure.

Our implementation has a graphic interface to display the results of the search. Various tolerances, such as errors in point locations, errors in distance measurements, and minimum subset sizes can be specified. Multiple solutions can be displayed simultaneously, and the user can change parameters and view the results in real time. It should be noted that we use actual sets of drug molecules with known pharmacophoric subsets, and the above procedure is quite effective in determining the *correct* solution (as detected by chemists).

4.2 Extension to Multiple Molecules

Previous attacks on the problem often use a clique-finding approach similar to the above to determine common subsets. This is then extended to multiple molecules in a natural way, where one molecule is compared with each of the rest simultaneously, and then the procedure is iterated [24]. This approach is simple, but computationally expensive, and cannot handle situations where a pharmacophore may exist in most, but not all, molecules. It should also be noted that these methods do not address the case where molecules can be in one of many conformations.

In general, it is easier to determine whether a given substructure exists in a conformation than to search for some common substructure that exists in two conformations. Therefore, our approach in the case of multiple molecules uses two primitives:

P1 that takes two sets of conformations and produces a set of common substructures from them, and

P2 that takes such a set of substructures, and another molecule, and *filters* the invariants through this molecule, retaining only those that are contained in it.

One simple strategy would choose two molecules arbitrarily, apply P1 on them, and use P2 repeatedly on the result and the rest of the molecules. Another possible strategy would be to view common substructures as sets of *pseudo-conformations*, and extend P1 to process such pseudo-conformations as well. Currently, we are investigating the efficacy of such strategies on various actual sets of drug molecules.

It is important to represent molecules and invariant groups compactly, to make P1 and P2 more efficient. To do this, we use techniques similar in spirit to geometric hashing [33], a technique used for pattern recognition in computer vision, and recently also in computational biology [27]. We hash all the conformations of all the molecules into a table, using a hash function of a geometric nature defined on pairs or triplets of atoms, and taking all such subsets. With every such table entry, we associate the conformation which contributed to a hit on it. Now, if we hash the pharmacophores in a similar fashion, we can identify which conformations are hit frequently, and so can deduce which molecules are likely to possess at least one conformation containing the pharmacophore.

References

1. T. Akutsu, M. Halldorsson, On the approximation of largest common point sets, *Proc. of the International Symposium on Algorithms and Computations, Springer-Verlag Lecture Notes on Computer Science* **834** (1987), pp. 405–413.
2. H. Alt, K. Melhorn K., H. Wagener and E. Welzl, Congruence, Similarity, and Symmetries of Geometric objects, *Discrete Comput. Geom.* **3** (1988), pp. 237–256.
3. L.M. Balbes, S.W. Mascarella, D.B. Boyd, A perspective of modern methods for computer aided drug design, *Reviews in Computational Chemistry* **5** (1994), VCH Publishers Inc., pp. 337–379.
4. D.B. Boyd, Compendium of molecular modeling software, *Reviews in Computational Chemistry* **4** (1993), VCH Publishers Inc., pp. 229–257.
5. R.P. Brent, *Algorithms for finding zeros and extrema of functions without calculating derivatives*, Ph.D. Thesis, Stanford University, 1971.

6. Brint, A. T. and Willet P, Algorithms for the identification of three-dimensional maximal common substructures. *J. Chem. Inf. Comput. Sci.* **27** (1987), pp. 152–158.

7. C.E. Bugg, W.M. Carson, and J.A. Montgomery, Drugs by design, *Scientific American*, December 1993, pp. 92–98.

8. D.E. Clark, G. Jones, P. Willet, P.W. Kenny, and R.C. Glen, Pharmacophoric pattern matching in files of three-dimensional chemical structures: Comparison of conformational searching algorithms for flexible searching, *J. Chem. Inf. Comput. Sci.* **34** (1994), pp. 197–206.

9. M.L. Connolly, Solvent-accessible surfaces of proteins and nucleic acids, *Science* **221** (1983), pp. 709–713.

10. M.L. Connolly, Analytical molecular surface calculation, *J. of Applied Crystallography* **16** (1983), pp. 548–558.

11. H. Edelsbrunner, M. Facello, P. Fu, and J. Liang, Measuring proteins and voids in proteins, *Technical Report*, HKUST-CS94-19, Department of Computer Science, Hong Kong University of Science and Technology, 1994.

12. R.C. Glen, G.R. Martin, A.P. Hill, R.M. Hyde, P.M. Woollard, J. Salmon, J. Buckingham and A. Robertson, Computer-aided design and synthesis of 5-substituted Tryptamins and their pharmacology at the 5-HT Receptor: discovery of compounds with potential anti-migraine properties, *J. Med. Chem.*, **38** (1995), pp. 3566-3580.

13. L.J. Guibas and J. Stolfi, Primitives for the manipulation of general subdivisions and the computation of Voronoi diagrams, *ACM Transactions on Graphics*, **4** (1985), pp. 74–123.

14. D. Gusfield and R. W. Irving, *The stable marriage problem: structure and algorithms*. MIT Press, Cambridge, 1989.

15. D. Halperin, J.-C. Latombe and R. Motwani, Dynamic maintenance of kinematic structures, Manuscript, 1996.

16. D. Halperin and M.H. Overmars, Spheres, molecules, and hidden surface removal, *Proc. 10th ACM Symposium on Computational Geometry*, Stony Brook, 1994, pp. 113–122.

17. L.E. Kavraki, *Random networks in configuration space for fast path planning*, Ph.D. Thesis, Stanford, 1995.

18. L. Kaufman and P.J. Rousseeuw, *Finding groups in data an introduction to cluster analysis*, Wiley, NY, 1990.

19. S. Khanna, R. Motwani, and Frances F. Yao, Approximation algorithms for the largest common subtree problem, Report No. STAN-CS-95-1545, Department of Computer Science, Stanford University (1995).

20. G. Klebe and T. Mietzner, A fast and efficient method to generate biologically relevant conformations, *J. of Computer Aided Molecular Design* **8** (1994), pp. 583-606.

21. A.R. Leach, A survey of methods for searching the conformational space of small and medium size molecules, *Reviews in Computational Chemistry* **2** (1991), VCH Publishers Inc., pp. 1–55.

22. T. Lengauer, Algorithmic research problems in molecular bioinformatics, *IEEE Proc. of the 2nd Israeli Symposium on the Theory of Computing and Systems*, 1993, pp. 177–192.

23. B. Lee and F.M. Richards, The interpretation of protein structure: Estimation of static accessibility, *J. of Molecular Biology* **55** (1971), pp. 379–400.

24. Y.C. Martin, M.G. Bures, E.A. Danaher, J. DeLazzer, I. Lico, and P.A. Pavlik, A fast new approach to pharmacophore mapping and its application to dopaminergic and benzodiazepine agonists, *J. of Computer-Aided Molecular Design* **7** (1993), pp. 83–102.

25. P.G. Mezey, Molecular surfaces, in *Reviews in Computational Chemistry*, Vol. I, K.B. Lipkowitz and D.B. Boyd, Eds., VCH Publishers, 1990, pp. 265–294.

26. K. Mulmuley, *Computational Geometry: An Introduction Through Randomized Algorithms*, Prentice Hall, New York, 1993.

27. R. Norel, D. Fischer, H.J. Wolfson, and R. Nussinov, Molecular surface recognition by a computer vision-based technique, *Protein Engineering* **7** (1994), pp. 39–46.

28. F.M. Richards, Areas, volumes, packing, and protein structure, in *Annual Reviews of Biophysics and Bioengineering* **6** (1977), pp. 151–176.

29. D.A. Pierre, *Optimization theory with applications*, Dover, NY, 1986.

30. A. Smellie, S.D. Kahn, and S.L. Tieg, Analysis of conformational coverage: 1. Validation and estimation of coverage, *J. Chem. Inf. Comput. Sci*, 35(1995), pp. 285–294.

31. Y. Takahashi, Y. Satoh and S. Sasaki, Recognition of largest common structural fragment among a variety of chemical structures, *Analytical Sciences* **3** (1987), pp. 23–28.

32. Tripos Associates Inc., Sybyl Manual, St. Louis, MO.

33. H. J. Wolfson, Model-based object recognition by geometric hashing, *Proc. of the 1st European Conference on Computer Vision* (1990), pp. 526-536.

Ray-Representation Formalism for Geometric Computations on Protein Solid Models

Michael G. Prisant[1]

Departments of Chemistry and Computer Science
Duke University
Durham, N.C. 27708

Abstract. Ray-representation or ray-rep formalism provides a comprehensive and simple approach for geometric computations on molecular solid models. These methods allow model formulation in terms of simple constructive solid geometry. In particular, the van der Waals exclusion volume is described by computing the ray-representation of a union of spheres and the solvent exclusion volume is computed by Minkowski dilation and erosion. Volume and area properties are calculated, respectively, by a "pile-of-bricks" and "collocation-of-tiles" interpretation of the ray-rep. Labeling the chemical character of surface patches is facilitated by the intrinsic point ordering of the ray-rep. Definition of internal cavities can be accomplished by equivalence-set clustering of internal void segments. Finally, a Boolean intersection procedure determines placement of crystallographic waters with respect to the protein solvent excluded volume.

1   Introduction

Space filling models are often invoked to rationalize the chemical behavior of proteins. Physio-chemical properties treated by geometric methods include: packing density,[1, 2] folding energetics,[3, 4, 5] and molecular recognition.[6]

   We have developed new methods – based on the idea of the ray representation – for manipulating molecular shapes and computing their geometric properties. Ray-representation calculations combine speed, accuracy, and extraordinary conceptual simplicity. The ray representation formalism provides a unified means of dealing with geometric issues in molecular modeling.

2   Background on the Ray Representation

In constructive solid geometry, a complex solid object can be represented as a list of simpler primitive objects, their positions, their material identity, and instructions on how the objects are to be put together. The primitive objects – notable examples are spheres, ellipsoids or blocks – can be analytically described by mathematical equations which define the boundaries of their solid volume. The instructions describing how to construct the complex object from the simpler primitives can be given by Boolean union, intersection, and difference operations.[7, 8]

Ray-casting and ray-reps can greatly simplify practical use of constructive solid geometry. This methodology provides a discretized quasi one-dimensional representation for what at first seems to be a continuous three-dimensional problem. The process of ray-casting imitates the passage of light through the continuous representation of a solid object. The rays of light are modeled by a grid of parallel straight lines. During the ray casting procedure, we compute where each of these lines enters and exits the solid object. The ray representation of the solid object is given by the list of ray entry and exit point locations. Ray-reps of arbitrary shapes can be calculated in a straightforward manner. This is because it has been shown that the ray-rep of a complex solid can be obtained by successive Boolean operations on the ray-reps of the individual primitives.[9, 10] In this manner, the ray-rep of a composite solid can be computed on a ray by ray basis. An example of this type of construction is shown on the left side of Fig. 1. A two dimensional torus is constructed by the ray by ray Boolean difference of a large and small ellipse.

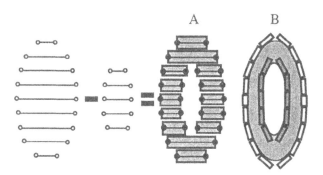

Fig. 1. **LEFT:** Construction of the ray rep of a composite object from simpler primitives. The torus ray-rep is computed by taking the ray by ray Boolean difference of two ellipses. **RIGHT:** Examples of unsophisticated algorithms for calculating volume and surface area. A) "Pile-of-Bricks" volume calculation and B) "Collocation of Tiles" area calculation.

It has been widely thought that the conceptual benefits of ray-reps were balanced by high computational cost. This balance of perceived benefit and cost motivated the construction of special purpose computers for ray-casting calculations.[11] Initial work on protein ray-casting in my laboratory by Martha Head was accomplished on just such a machine constructed by Gershon Kedem.[12] These studies laid out general algorithmic approaches and provided "proof-of-principle". Subsequent work, to be described here, has demonstrated that molecular ray-casting – which is mostly concerned with sphere unions – can be simply and rapidly accomplished on general purpose architectures.

Early machine-specific implementations of the ray-representation were limited to a short material tag with each ray entry point. In our previous minimal implementations, the tag was only able hold an atom serial number. This allowed specification of the primitive which gave rise to a particular point. In our present implementation, an extended tag is provided for both entry and exit points. This extended tag also stores other information derived from the parenting primitive including a surface normal and differential area.

Calculation of physical properties from extended tag ray representations is conceptually simple. Consider two naive algorithms for volume and surface area of a compound object using the ray representation. In the *pile-of-bricks* volume calculation shown on the right side of Fig. 1A, the lengths of all line segments connecting ray entry and exit points are computed. The scaled sum of these lengths estimates the volume of the compound object. In the *collocation-of-tiles* area calculation shown in Figure 1B, an area associated with each entry and exit points is derived from the placement of the point on the parenting primitive. The sum of these tile areas estimates the surface area of the compound object. In a similar fashion, visualization of the final ray-rep is straightforward because a surface normal is associated with each point and the points are already sorted front to rear.

3   Applications to Molecular Solid Geometry

3.1   The Fused Sphere Model

The fused sphere model represents a protein molecule as a union of spheres. Each atom in the molecule is taken to be a hard sphere of given van der Waals radius. The sphere centers are placed so as to coincide with atomic centers in the molecule. The atomic van der Waals radii – which determine the radius for non-bonded interactions – are typically larger than the covalent radii which set the distance between bonded adjacent spheres. Hence the spheres in this representation overlap and appear fused. Fused-sphere models define a spatial boundary of a molecule or *van der Waals exclusion volume* with respect to non-bonding interactions. The spirit of this model is that the spheres of neighboring non-bonding atoms should not interpenetrate. Thus, this spatial boundary is, in essence, an energy surface for non-bonding interactions within the protein and with other molecules *in the absence of solvent.*

The straightforward algorithms presented in the previous section are the starting point for simple, accurate, and rapid calculations. We have written a short *C* program ($<$ 300 lines of code) to implement the ray-casting calculations. Program accuracy and timing are shown in Fig. 2 as a function of casting density for myoglobin, a 1261 atom protein. "Exact" values for the area and volume of this system were calculated using the analytical treatment of Dodd and Theodorou[13] and then compared to the ray-casting result. The bottom panel provides timing information for the calculations on a 133 Mhz Pentium processor. Code execution time is linear with respect to the number of *non-empty rays*

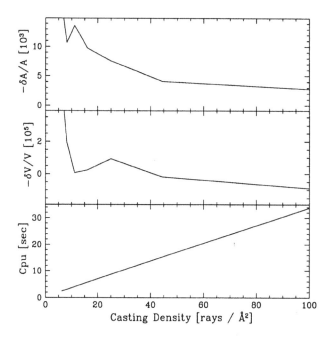

Fig. 2. Ray representation calculations of solvent accessible surface as function of grid density for myoglobin. The top, middle, and bottom panels display respectively: Collocation of tiles area calculation, pile of bricks volume calculation, and single processor computation times.

and the average number of atom spheres intersected by each ray. The number of non-empty rays is proportional to $N^{\frac{2}{3}} \times \rho$ where N is the number of atoms and ρ is the casting density in $rays /Å^2$. Because protein density is more or less constant, the average number of atom spheres intersected by each ray is independent of protein size. At a casting density of 25 $rays/Å^2$, the ray-rep computation takes approximately 10 seconds. As can be seen in the middle and top panels this casting density provides excellent accuracy in volume estimation – better than one part in 10^5 – and modest accuracy in area estimation – better than one part in 10^2.

Simple parallelization of the calculation can be accomplished by partitioning of casting space. Calculation time for myoglobin at 100 $rays/Å^2$ on a 4 processor 90 MHz Ross HyperSparc platform is just under 10 seconds or 3.5 fold faster than the corresponding single processor execution time.

3.2 Computation of Molecular Contact Surface by Minkowski Procedure

The geometric construct of a surface boundary through which solvent cannot penetrate or *solvent exclusion volume* is key to many descriptions of how a protein interacts with surrounding water solvent. The surface of the fused sphere solid model of the protein does not correspond to this outer solvent accessible boundary. This is because the solvent is of finite size and therefore cannot come into contact with all portions of the fused sphere solid model's reentrant surface. The molecular contact surface defining a protein solvent exclusion volume was first described by Richards [14] and computed by Connolly.[15, 16] Since then, computation of molecular contact surfaces has received extensive attention from diverse sources.[17, 18, 19, 20, 21, 22] However, we believe that ray-rep techniques provide a much simpler approach to this problem than other methods. The algorithm to be described can be implemented in approximately 200 lines of C code.

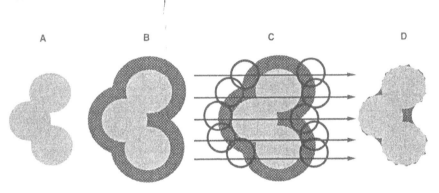

Fig. 3. Pictorial depiction of the Minkowski operation: A) Fused VDW spheres are the starting point; B) The spheres are bloated by the solvent radius; C) The bloated object is ray-cast and a casing of solvent radius spheres are positioned at the ray entry and exit points; D) The casing spheres are subtracted from the bloated object.

We now discuss how the ray-representation can be used to greatly simplify calculation of the solvent exclusion surface of proteins. Fig. 3 shows the dilation and erosion steps in the process.[23] The procedure begins with the original collocation of fused van der Waals spheres. These atomic spheres are bloated by a radius corresponding to a probe sphere of water.[2] The bloating of these spheres by the radius of the probe or solvent molecule fills in all crevices which cannot accommodate a solvent molecules. Next, the bloated object is ray-cast and a casing of solvent spheres is located at the ray entrance and exit points. An compound

---

[2] The bloating step is equivalent to unioning the original fused sphere model with a casing of probe spheres positioned at all of its surface points.

object corresponding to the union of these casing spheres is constructed. Finally, a *new* ray representation is computed by using Boolean calculus to subtract the probe sphere casing from the bloated protein. This subtraction has the effect of blending reentrant concave surfaces on the protein according to the probe size while recovering the convex surfaces of the original unbloated fused sphere model. The whole procedure – defined by addition and subtraction of spheres – is a special case of a geometric ansatz defined by Minkowski. [9, 10, 24, 25] The surface of the final blended object corresponds to a shrink-wrapped skin around the original protein in which all reentrant surfaces inaccessible to the solvent probe are smoothed over. This surface represents the true solvent accessible or molecular contact surface and is fully equivalent that defined in Connolly's work.

Computation times for the Minkowski procedure have a linear dependence on the number of segments to be subtracted. This number of segments equals $n \times \rho r^2$ or the number of erosion balls times the number of rays per erosion ball. The number of erosion balls, n, is proportional to $N^{\frac{2}{3}}\rho$ or number of non-empty rays. The number of rays per erosion ball, ρr^2, depends in turn on casting density times the square of the probe radius. Thus calculation times have a quadratic dependence on casting density. Single processor execution times for myoglobin on a Pentium 133 range from 0.730 seconds at 4 $rays/\mathring{A}^2$ to 378.300 seconds at 100 $rays/\mathring{A}^2$. Volume calculations attain better than 5 parts per thousand convergence at 25 $rays/\mathring{A}^2$ and area calculations at 44.5 $rays/\mathring{A}^2$.

Program parallelization of these calculations via casting space partitioning yields a linear speedup with the number of processors. Execution on a 4 processor 90 MHz Ross HyperSparc platform, as an example, takes 122 seconds for myoglobin at 100 $rays/\mathring{A}^2$.

3.3   Chemical Character of Molecular Contact Surface

We would like to label all points on the molecular contact or blended surface with regard to (i) their chemical identity and (ii) whether they lie on reentrant locally concave or non-reentrant locally convex portions of the surface. Ray-representation formalism greatly facilitates this labeling.

Fig.4 shows how both chemical identity and local curvature are ascertained. We can compare the locus of points in the blended and unblended ray-representation. Coincident points – within some distance criterion ϵ – on both surfaces are labeled *associable*. Because these points are coincident, their chemical identity can simply be taken from the corresponding point on the original unblended or fused sphere ray-rep. We also remark that these points always lie on non-reentrant or locally convex pieces of surface. Conversely, points on the blended surface which do not coincide with a point on the fused sphere surface are labeled *non-associable*. Such points always lie on a reentrant or locally concave portions of surface. The chemical identity of such a point can be assigned by finding its closest neighbor on the fused sphere surface.

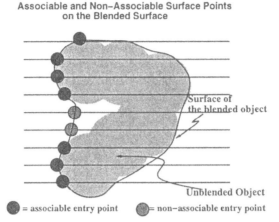

Fig. 4. Distinction between associable and non-associable entry points on the blended surface.

Fig. 5. A) Ray traversing one pair of entry and exit points contains solid space. B) Ray traversing more than one pair of entry and exit points contains internal free spaces.

3.4 Captured Cavities and their Properties

The Minkowski surface of a protein defines the protein's boundary and surface area with respect to solvent. We are interested in learning about cavities in the protein contained by this Minkowski surface. If totally enclosed, we might imagine that these cavities are important structural elements in protein folding.

How can we use the ray representation to learn more about enclosed cavities in the protein interior? The simplicity of the ray-rep based algorithm is that it basically creates associated groupings of void segments. These groupings of void segments provide a ray representation of the individual cavities. In this approach, the shape and chemical identity of cavity surface points are rigorously and unambiguously defined.

As shown in Fig. 5, individual rays can be classified by the number of they enter and exit an object. If the ray has 0 sets of entry and exit points, then it is **empty**. If the ray has 1 set of entry and exit points then it is **solid**. If the ray has multiple sets of entry and exit points, then it contains **void** segments. The number of void segments equals the number of entry-exit pairs minus one. The lead of each void segment is defined by the exit of the previous solid segment.

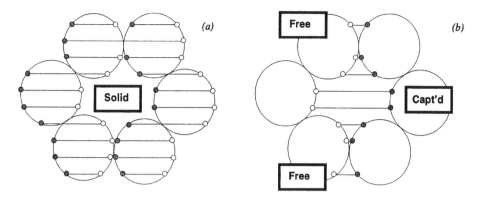

Fig. 6. (a) Solid segments. (b) Void segments are either captured or free.

The trail is defined by the entry of the succeeding solid segment.

As shown in Fig. 6, void segments can be further classified into two categories according to the neighborhood of the ray to which they belong. Void segments which overlap empty space on adjoining rays are categorized as *free* void segments. Examples of free void segments can be seen in the re-entrant pockets of the sphere ring depicted in the figure. Conversely, void segments which do not overlap empty space are categorized as *captured* segments. These segments can be seen in the interior of the ring of spheres.

We are now ready to map out the enclosed and open caverns in the protein. Algorithmically this means that we wish to consider the list of void segments within the protein. We wish to divide this list into associated groups or equivalence sets. Segments which are connected through a series of overlaps with neighboring segments can be grouped together using standard clustering algorithms[26] and labeled in the manner described by Hoshen and Kopelman.[27] Each equivalence set of void segments defines a "cavern" in the protein. Now, if any one void segment in an equivalence set overlaps empty space, then that cavern opens to the outside. Conversely if no one void segment in an equivalence set overlaps empty space, then that cavern is completely enclosed within the protein.

3.5 Identification of Internal Waters

Identification of internal waters in crystallographic structures is an important element of understanding the role of water in stabilizing protein structures. Traditional evaluation of the internal character of water was decided on the basis of metric geometry or the number of hydrogen bonds it could form with internal protein atoms. Recently, Varadarajan and Richards developed a numerical methodology based on the use of probe spheres.[28] However this methodology effectively uses the *original* Richards definition of solvent accessible surface as the boundary of the protein. Its criteria for classifying waters as internal is there-

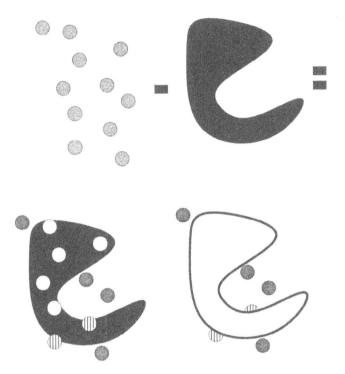

Fig. 7. Boolean solid difference operation allows unambiguous identification of internal waters. Waters which disappear during the difference operation – here shown as white disks – are completely inside the solvent excluded volume defined by the protein. Waters which are unchanged by the diffference operation – here shown as grey disks – are completely outside the solvent excluded volume. Waters which do not disappear but whose solid shape is altered – here shown as disks with vertical lines – straddle the molecular contact surface.

fore not fully consistent with the definition of the protein's solvent accessible boundary in terms of the Connolly or molecular contact surface.

For a given space-filling model of protein structure, the ray-rep allows unambiguous determination of buried waters. The basic idea is to make use of a solid Boolean difference scheme. This scheme is illustrated in Fig. 7. The following calculation steps are required. First, we compute the ray representation of the blended protein with all captured internal cavities filled. This can be accomplished by Boolean addition of the ray-representation of the internal cavities with the blended protein. Second, we compute the unblended ray representation of all crytallographically located waters. Third, we compute the ray representation corresponding to the Boolean difference between these two sets. The tags in the ray representation which results from this difference operation provide a list of remaining waters. Waters which are fully inside the protein's molecular contact surface will be absent from this list.

4 Summary, Discussion, and Prospects

This paper has developed tools for using ray representation in the geometric analysis of space-filling models of proteins. The emphasis here has been on volumetric aspects or more precisely geometric properties which could be treated within the framework of the "pile-of-bricks" interpretation of the ray representation. Techniques for the treatment of the following aspects of protein structure were developed: definition of fused sphere models, definition of molecular contact surface, description of cavities and grooves, and determination of internal waters,

Many of the geometric issues discussed in this paper have been treated – at least in part – by other techniques. However, we argue that the ray representation provides a unifying and simplifying foundation for geometric calculations in proteins. Geometric analysis on the basis of ray-representations is clearly defined and essentially unambiguous. The simplifying aspect of the ray representation is especially seen in Minkowski treatment of molecular contact surface, the determination of captured void cavities, and the computation of volume. This simplification means that algorithmic ambiguities need not muddy scientific interpretation of calculation results. In fact, the complexity of other techniques has discouraged routine computation of many of the quantities clearly defined here. Given this success with basic geometric properties, we believe that ray representations have the potential to be of considerable help in molecular docking problems and structure based rational drug design.

Acknowledgements

We thank Sun Microsystems and the Lord Foundation for their financial support. Our first protein ray casting calculations were performed by my graduate student Martha Head on the Duke Ray-Casting Engine or RCE 1.0, a special purpose ray-casting computer. These studies laid the groundwork for the present calculations using generic C code on general purpose architectures. We thank Professor Gershon Kedem of the Duke University Department of Computer Science for making this machine available for protein calculations.

References

1. F. M. Richards. Folded and unfolded proteins: An introduction. In Thomas E. Creighton, editor, *Protein Folding*, pages 1–58. W. H. Freeman and Company, New York, 1992.
2. J. L. Finney. Volume occupation, environment and accessibility in proteins, the problem of the protein surface. *J. Mol. Biol.*, 96:721–732, 1975.
3. Cyrus Chothia. The nature of the accessible and buried surfaces in proteins. *J. Mol. Biol.*, 105:1–14, 1976.
4. C. Frommel. The apolar surface area of amino acids and its empirical correlation with hydrophobic free energy. *J. Theor. Biol.*, 111:247–260, 1984.

5. D. Eisenberg and A. D. McLachlan. Solvation energy in protein folding and binding. *Nature*, 319:199–203, 1986.

6. Irwin D. Kuntz. Structure-based strategies for drug design and discovery. *Science*, 257:1078–1082, 1992.

7. A. A. G Requicha. Mathematical models of solid objects. Technical Report Technical Memorandum 28, Production Automation Project, University of Rochester, Available from CPA/COMEPP, Cornell University, Ithaca, NY 14853-7501, November 1977.

8. A. A. G Requicha and R. B. Tilove. Mathematical foundations of constructive solid geometry: general topology of closed regular sets. Technical Report Technical Memorandum 27, Production Automation Project, University of Rochester, Available from CPA/COMEPP, Cornell University, Ithaca, NY 14853-7501, June 1978.

9. J. P. Menon, R. J. Marisa, and J. Zagajac. More powerful solid modeling through ray representations. Technical Report CPA92-4, Sibley School of Mechanical and Aerospace Engineering, Cornell University, Ithaca, NY 14853-7501, 1992.

10. J. P. Menon and H. B. Voelcker. Mathematical foundations: set theoretic properties of ray representations and minkowski operations on solids. Technical Report CPA92-9, Sibley School of Mechanical and Aerospace Engineering, Cornell University, Ithaca, NY 14853-7501, January 1992.

11. John Ellis, Gershon Kedem, Richard Marisa, Jai Menon, and Herbert Voelcker. Breaking barriers in solid modeling. *Mechanical Engineering*, pages 28–34, February 1991.

12. M. S. Head, G. Kedem, and M. G. Prisant. Application of the ray representation and a massively parallel special purpose computer to problems of protein structure and function: I. Methodology for calculation of molecular contact surface, volume and internal free space. Technical Report CS-1994-31, Department of Computer Science, Duke University, 1994.

13. Lawrence R. Dodd and Doros N. Theodorou. Analytical treatment of the volume and surface area of molecules formed by an arbitrary collection of unequal spheres intersected by planes. *Mol. Phys.*, 72(6):1313–1345, 1991.

14. Frederic M. Richards. Areas, volumes, packing, and protein structure. *Ann. Rev. Biophys. Bioeng.*, 6:151–176, 1977.

15. Michael L. Connolly. Solvent-accessible surfaces of proteins and nucleic acids. *Science*, 221(4612):709–713, August 1983.

16. Michael L. Connolly. Analytical molecular surface calculation. *J. Appl. Crystallogr.*, 16:548–558, 1983.

17. Michael L. Connolly. The molecular surface package. *J. Mol. Graphics*, 11:139–141, 1993.

18. Juan Luis Pascual-Ahuir and Estanislao Silla. GEPOL: An improved description of molecular surface. I. building the spherical surface set. *J. Comp. Chem.*, 11(9):1047–1060, 1990.

19. W. Heiden, T. Goetze, and J. Brickmann. Fast generation of molecular surfaces from 3D data fields with an enhanced "marching cube" algorithm. *J. Comp. Chem.*, 14(2):246–250, 1993.

20. Michel Sanner, Arthur J. Olson, and Jean Claude Spehner. Fast and robust computation of molecular surfaces. In *Proc. 11th ACM Symp. Comp. Geom, C6-C7*. ACM, 1995.

21. H. Edelsbrunner and E. Mucke. Three-dimensional alpha shapes. *ACM Transactions on Graphics*, 13(1):43–72, 1994.

22. Amitabh Varshney, Frederick P. Brooks Jr., and William V. Wright. Linearly scalable computation of smooth molecular surfaces. *IEEE Computer Graphics and Applications*, 14(5):19–25, 1994.

23. H. J. A. Heijmans and C. Ronse. The algebraic basis of mathematical morphology: I. dilations and erosions. *Computer Vision, Graphics, and Image Processing*, 50:245–295, 1990.

24. Jean-Paul Serra. *Image Analysis and Mathematical Morphology*. Academic Press, New York, 1982.

25. Benoit E. Mandelbroit. *The Fractal Geometry of Nature*. Freeman, New York, 1983.

26. Dietrich Stauffer and Amnon Aharony. *Introduction to Percolation Theory*. Taylor and Francis, Washington, DC, 1992.

27. J. Hoshen and R. Kopelman. Percolation and cluster distribution. I. cluster multiple labeling technique and critical concentration algorithm. *Phys. Rev. B.*, 14(8):3438–3435, 1976.

28. R. Varadarajan and F. M. Richards. Crystallographic structures of ribonuclease s variants with nonpolar substitution at position 13: Packing and cavities. *Biochemistry*, 31:12315–12327, 1992.

Column-Based Strip Packing Using Ordered and Compliant Containment

Karen Daniels[*][1] and Victor J. Milenkovic[**][2]

[1] Harvard University and University of Miami
[2] University of Miami

Abstract. The *oriented strip packing* problem is very important to manufacturing industries: given a strip of fixed width and a set of many (> 100) nonconvex polygons with 1, 2, 4, or 8 orientations permitted for each polygon, find a set of translations and orientations for the polygons that places them without overlapping into the strip of minimum length. Heuristics are given for two versions of strip packing: 1) translation-only and 2) oriented. The first heuristic uses an algorithm we have previously developed for translational containment: given polygons $P_1, P_2, \ldots, P_k$ and a fixed container C, find translations for the polygons that place them into C without overlapping. The containment algorithm is practical for $k \leq 10$. Two new containment algorithms are presented for use in the second packing heuristic. The first, an *ordered containment* algorithm, solves containment in time which is only linear in k when the polygons are a) "long" with respect to one dimension of the container and b) ordered with respect to the other dimension. The second algorithm solves *compliant containment*: given polygons $P_1, P_2, \ldots, P_{l+k}$ and a container C such that polygons $P_1, P_2, \ldots, P_l$ are already placed into C, find translations for $P_{l+1}, P_{l+2}, \ldots, P_{l+k}$ and a nonoverlapping translational *motion* of $P_1, P_2, \ldots, P_l$ that allows all $l + k$ polygons to fit into the container without overlapping.

The performance of the heuristics is compared to the performance of commercial software and/or human experts. The results demonstrate that fast containment algorithms for modest values of k ($k \leq 10$) are very useful in the development of heuristics for oriented strip packing of many ($k >> 10$) polygons.

---

[*] Harvard University, Division of Applied Sciences, and University of Miami. Email: daniels@das.harvard.edu. This research was funded by the Textile/Clothing Technology Corporation from funds awarded by the Alfred P. Sloan Foundation and by NSF grants CCR-91-157993 and CCR-90-09272.

[**] University of Miami, Department of Math and Computer Science. Email: vjm@cs.miami.edu. This research was funded by the Textile/Clothing Technology Corporation from funds awarded to them by the Alfred P. Sloan Foundation, by NSF grant CCR-91-157993, and by a subcontract of a National Textile Center grant to Auburn University, Department of Consumer Affairs.

1 Introduction

Many industries fabricate parts by cutting them from stock material. Often, stock material comes in rolls or sheets of fixed width and indeterminate length. Such industries need to solve the *strip packing* problem: pack a set of polygons onto a strip of fixed width and minimum length. If the material has a grain then usually each polygon is permitted only a fixed number (1, 2, 4, or 8) of orientations. Figure 1 shows an example of oriented strip packing from the apparel industry. Each part has four possible orientations except for the rectangles which have either one or two.[3] These layouts can have hundreds of parts, and oriented strip packing is NP-complete. Most layouts are currently done by a human using a CAD/CAM system, and the labor cost and the cost of wasted material are considerable.

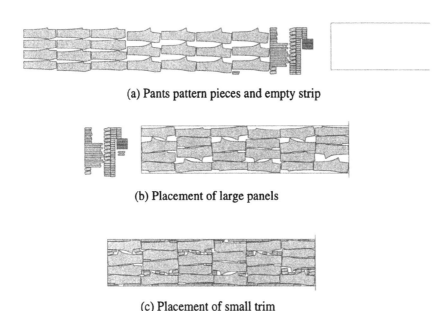

(a) Pants pattern pieces and empty strip

(b) Placement of large panels

(c) Placement of small trim

Fig. 1. An oriented strip packing task from the apparel industry

Recently we have developed fast algorithms for translational *containment* [1, 16, 2, 13, 4, 3, 15]: given nonconvex polygons $P_1, P_2, \ldots, P_k$ and a nonconvex polygonal container C, find translations for the polygons that pack them into C. For a modest number of polygons ($k \leq 10$), our containment algorithms find a solution or determine that none exists in a few minutes on a desk top workstation.

---

[3] For some parts, a small *tilt* is also permitted (typically up to 3 degrees). However, we deal with tilt in a post-processing *compaction* [11, 18, 12] phase.

Unless P = NP, one probably cannot hope to generalize this algorithm in a way that will be practical for $k_i > 100$ and/or multiple orientations. However, it is our contention that a fast translational containment algorithm for modest k is a very useful tool for the development of good heuristics for oriented strip packing.

This paper gives heuristics for strip packing which are based on containment and are domain-independent. We first give a heuristic for translation-only strip packing. This heuristic is based on packing one column at a time. We found that to generalize it to oriented strip packing, we needed two extensions of our containment algorithm. The first, *ordered containment*, solves containment very quickly (running time depends only linearly on k) when the polygons are "long" with respect to one dimension of the container (the column) and ordered with respect to the other dimension. The second, *compliant containment*, packs k additional polygons into a container already containing l polygons. The l polygons are permitted nonoverlapping motion but are forbidden to "jump" over each other. Using these two new containment algorithms, we give a heuristic for oriented strip packing.

1.1 Related Work

Strip packing and other forms of layout are well studied problems, and we refer the reader to surveys [1, 5, 6, 22]. The idea of column-based strip packing goes back to the 1960s. Gurel [9, 8, 10] used a graph-theoretical approach to place pieces in columns. In industrial practice, columns are sometimes referred to as "breaks". Our work on translational containment is described in [1, 16, 2, 13, 4, 3, 15] (see [1] for a survey of containment research).

Figure 1a illustrates the pieces required to manufacture twelve pairs of pants. In earlier work, we use a column-based method to place the large panel pieces (Figure 1b) [17], and then a containment-based method to place the smaller trim pieces[4] (Figure 1c) [1]. These methods generate pants layouts within the same production quality range as that imposed on human experts. The panel placement heuristic packs each column with the most "difficult" set of remaining panels. By assuming a particular x-translation for each panel, it can check each possibility very quickly. Ordered containment (Section 3) finds column packings without knowing the x-translations, and this permits us to generalize the "worst"-first strategy to a domain-independent heuristic. (See Section 7 for further comparison of the panel placement heuristic with our new work.)

For the most part, our work has focused on algorithms, not heuristics. Part of the reason for this emphasis is that it is difficult to measure the performance of heuristics objectively. However, recently a large domestic auto manufacturer challenged us with a large ($k = 50$) translation-only strip packing problem. Using our containment algorithm and a simple column-based heuristic (Section 2.3),

---

[4] Our pants layout software is part of our Automatic Marker Making Toolkit, which has been licensed by Gerber Garment Technologies. The toolkit also includes compaction, overlap reduction, and containment software.

we generated a solution with 83.94% cloth utilization.[5] By comparison, a naive first-fit strategy only achieved 81.71% utilization. The best commercial software achieved 85%, and the best human performance was 86.5%.[6]

1.2 Outline

The paper is organized as follows. Section 2 gives definitions and background relevant to containment, it presents the salient parts of our current containment algorithm, and then it describes a simple column-based heuristic for translational strip packing. Section 3 discusses ordered containment. Section 4 introduces compliant containment. Section 5 gives our column-based strip packing heuristic for polygons with multiple orientations, which relies on both ordered and compliant containment. Section 6 describes our implementation and gives results. Section 7 discusses future work.

2 Definitions, Background, and Translational Strip Packing

2.1 Definitions

The *Minkowski sum* [19, 7, 20, 21] of two point-sets (of $\mathbf{R}^2$ in the case of this paper) is defined

$$A \oplus B = \{a + b \mid a \in A,\ b \in B\}.$$

For a point-set A, let $\overline{A}$ denote the set complement of A and define $-A = \{-a \mid a \in A\}$. For a vector t, define $A + t = \{a + t \mid a \in A\}$. Note that $A + t = A \oplus \{t\}$.

A containment algorithm translates k polygonal regions $P_1, P_2, \ldots, P_k$ into a polygonal container C without overlap. If we denote $P_0 = \overline{C}$ to be the complement of the container region, then containment is equivalent to the *placement* of $k + 1$ polygons $P_0, P_1, P_2, \ldots, P_k$ in nonoverlapping positions. For translations t_i and t_j, $P_i + t_i$ and $P_j + t_j$ do not overlap if and only if $t_j - t_i \in U_{ij}$, where

$$U_{ij} = \overline{P_i \oplus -P_j}, \qquad 0 \le i, j \le k, i \ne j. \tag{1}$$

The set U_{ij} is the two-dimensional *configuration space* for placing P_j with respect to P_i. Clearly, $U_{ij} = -U_{ji}$. A *valid configuration* $\langle t_0 = (0,0), t_1, t_2, \ldots, t_k \rangle$ of $P_1, P_2, \ldots, P_k$ satisfies

$$t_j - t_i \in U_{ij}, \qquad 0 \le i < j \le k. \tag{2}$$

We set $t_0 = (0,0)$ because the container is not permitted to move. For the set of all valid configurations, we denote the range of $t_j - t_i$ by U_{ij}^*.

---

[5] Cloth utilization is the ratio of the sum of the areas of the placed polygons to the area of the strip in which they are placed, expressed as a percentage.

[6] These two results were for a slightly harder problem: the pieces had small buffer zones, and the cutting rules disallowed some types of piece contacts. To keep the challenge challenging, the company has asked us not to reveal the dataset.

2.2 Containment Algorithm

Our containment algorithm [16] takes as input a *hypothesis* $\mathcal{U} = \langle U_{ij}, 0 \leq i < j \leq k \rangle$. Equation 1 gives the initial values of the U_{ij}. (We sometimes denote these by U_{ij}^{init}.) The containment algorithm applies *restriction, evaluation,* and *subdivision* to the hypothesis. *Valid restriction* replaces some or all of the U_{ij} by subsets of themselves without losing any valid configurations for the hypothesis (Equation 2). *Evaluation* attempts to find at least one valid configuration for a hypothesis. *Subdivision* splits the hypothesis into two subhypotheses. The containment algorithm applies these three operations until it finds a solution or until all hypotheses are restricted to the empty set.

For purposes of this paper, we only need to describe restriction and subdivision. *Geometric restriction* replaces U_{ij} by $U_{ij} \cap (U_{ih} \oplus U_{hj})$, $0 \leq h \neq i \neq j \leq k$. *Linear programming (LP) restriction* computes the range R_{gh} of $t_h - t_g$ such that t_g and t_h are part of a valid configuration satisfying $t_j - t_i \in \text{CH}(U_{ij})$, $0 \leq i < j \leq k$, where $\text{CH}(U_{ij})$ is the convex hull of U_{ij}. It replaces U_{gh} by $U_{gh} \cap R_{gh}$. Our algorithms repeatedly apply restriction until a steady state is reached. Both geometric and LP restriction are valid restrictions.

Subdivision selects one U_{ij} in $\mathcal{U}$, splits U_{ij} into U_{ij}' and U_{ij}'', and creates subhypotheses $\mathcal{U}'$ and $\mathcal{U}''$ by substituting either U_{ij}' or U_{ij}'' for U_{ij}. If U_{ij} has multiple components, subdivision splits off one component. If not, it cuts U_{ij} with a line which is chosen using a *distance-based subdivision* technique [1, 16]. Distance-based subdivision ultimately cuts each U_{ij} into convex regions. If all U_{ij} in the current hypothesis $\mathcal{U}$ are convex, then the output of LP restriction is U_{ij}^* (the range of $t_j - t_i$ with respect to the valid configurations of $\mathcal{U}$).

2.3 Translational Strip Packing Heuristic

Given our containment algorithm, we can define a simple column-based translational strip packing heuristic. Suppose we are only willing to run our containment algorithm with $k \leq 5$ (since we have to call it many times). Order the polygons by diminishing length (x-dimension). To start a new column, set its length to that of the longest unplaced piece. Repeatedly attempt to place the next available polygon into the column using the procedure given below. If some polygon does not fit, skip to the next one. If no more polygons fit into a column, start a new column. The heuristic is column-greedy: once a column is placed, that column is never removed from the layout.

To place a polygon into the current column, *remove* the last four polygons placed in the column (or all of them if there are less than four) and then run containment to place them back in along with the new polygon. This is the heuristic we used on the auto manufacturer's challenge problem. The four steps back/five steps forward strategy permits the algorithm to correct "mistakes" in the placement of the last few polygons. This strategy generates a much tighter packing than a greedy one-polygon-at-a-time strategy.

3 Ordered Containment

Our column-based strip packing algorithm repeatedly solves containment for a column and a set of polygons to place in the column. Polygons are either "short" or "long" in the x-direction. "Short" polygons can be placed side by side in the column. "Long" polygons must be stacked in the column and cannot "pass" each other. Once the "long" polygons are placed, they have a particular *ordering* in the y-direction.

This section gives a rigorous definition of an *ordering* and a *strong ordering* for a set of polygons in a container. It then gives an *ordered containment* algorithm that finds a solution to containment for a strong ordering or reports that none exists. The running time of the algorithm depends on the complexity of the polygons, but it depends only linearly on k.

Using ordered containment, we can rapidly find ways of packing sets of "long" polygons into the container. We pack the short polygons using *compliant containment* which is described in Section 4.

3.1 Orderability

For $1 \leq i \neq j \leq k$, define $U_{ij}(C) = U_{ij}^{\mathrm{init}} \cap (U_{i0}^{\mathrm{init}} \oplus U_{0j}^{\mathrm{init}})$ which is U_{ij} after (geometric) restriction with respect to the container. We define P_i and P_j to be *orderable* ("long") when $U_{ij}(C)$ has at least two connected components $U_{ij}^{+}(C)$ and $U_{ij}^{-}(C)$ corresponding to the placement of P_j above or below P_i, respectively. If $U_{ij}(C)$ has more than two components, we can add the extra components to whichever of $U_{ij}^{+}(C)$ or $U_{ij}^{-}(C)$ is nearest. By definition, $U_{ij}^{+}(C) = U_{ji}^{-}(C)$. In practice, we only consider P_i and P_j to be orderable if we can partition the components using a horizontal line. The component(s) above the line are $U_{ij}^{+}(C)$ and the component(s) below the line are $U_{ij}^{-}(C)$.

An ordering is *strong* if for every ordered triple h, i, j such that P_i is between P_h and P_j, P_h cannot overlap P_j: $U_{hi}^{+}(C) \oplus U_{ij}^{+}(C) \subseteq U_{hj}^{+}(C)$.

3.2 Ordered Containment Algorithm

Suppose w.l.o.g. that $P_1, P_2, \ldots, P_k$ are strongly ordered in order of increasing index with respect to C. It takes $O(k^3)$ time to verify that an input ordering is strong (Section 6 describes a way to avoid this cost). This section gives an algorithm for solving containment for this ordering in time which is only linear in k. (Of course, it also depends on the number of vertices of $P_1, P_2, \ldots, P_k$ and C.) We begin by showing how to generate U_{0j}^{*}, $1 \leq j \leq k$, with respect to the ordering, in time which is only linear in k.

Lemma 1. *Given strongly ordered $P_1, P_2, \ldots, P_k$, U_{0j}^{*}, $1 \leq j \leq k$, with respect to that ordering can be generated in time which is, as a function of k, $O(k)$.*

Proof. Consider the following three loops, whose combined running time, as a function of k, is $O(k)$:

$$\textbf{For } i \leftarrow 1 \textbf{ to } k, \quad U_{0i} \leftarrow U_{0i}^{\text{init}}.$$
$$\textbf{For } i \leftarrow 2 \textbf{ to } k, \quad U_{0i} \leftarrow U_{0i} \cap (U_{0,i-1} \oplus U_{i-1,i}^{+}(C)).$$
$$\textbf{For } i \leftarrow k-1 \textbf{ to } 1, \, U_{0i} \leftarrow U_{0i} \cap (U_{0,i+1} \oplus U_{i+1,i}^{-}(C)).$$

Let $U_{0j}^{(2)}$ denote the value of U_{0j} after the second loop. We can prove by induction that $U_{0j}^{(2)}$ is *downwardly valid*: for each $t_j \in U_{0j}^{(2)}$, there exists a valid sub-configuration $\langle t_1, t_2, \ldots, t_j \rangle$ ($t_i \in U_{0i}$, $1 \leq i \leq j$, and $t_i - t_h \in U_{hi}^{+}(C)$, $1 \leq h < i \leq j$). This is vacuously true for $j = 1$. Suppose it is true for j. Clearly, $U_{0,j+1}^{(2)} \subseteq U_{0j}^{(2)} \oplus U_{j,j+1}^{+}(C)$. Hence, for each $t_{j+1} \in U_{0,j+1}^{(2)}$, there exists $t_j \in U_{0j}^{(2)}$ such that $t_{j+1} - t_j \in U_{j,j+1}^{+}(C)$. Choose a valid sub-configuration $\langle t_1, t_2, \ldots, t_j \rangle$. For $1 \leq h < j$, $t_{j+1} - t_h = t_j - t_h + t_{j+1} - t_j \in U_{hj}^{+}(C) \oplus U_{j,j+1}^{+}(C) \subseteq U_{h,j+1}^{+}(C)$, by the definition of strong ordering. Therefore $\langle t_1, t_2, \ldots, t_{j+1} \rangle$ is a valid sub-configuration, and $U_{0,j+1}^{(2)}$ is downwardly valid. Similarly, we can define *upwardly valid* and analyze the third loop. The output U_{0j} of the algorithm is both upwardly and downwardly valid and a subset of U_{0j}^{init}. Therefore, for each $t_j \in U_{0j}$, there exists a valid configuration $\langle t_1, t_2, \ldots, t_k \rangle$. Now we argue that no configurations which are valid with respect to the strong ordering are eliminated. The geometric restriction $U_{ij} \leftarrow U_{ij} \cap (U_{ih} \oplus U_{hj})$, $0 \leq h \neq i \neq j \leq k$ is a valid restriction [1, 3]. For this reason, the operations $U_{0i} \leftarrow U_{0i} \cap (U_{0,i-1} \oplus U_{i-1,i}^{+}(C))$ and $U_{0i} \leftarrow U_{0i} \cap (U_{0,i+1} \oplus U_{i+1,i}^{-}(C))$ in the second and third loops, respectively, are both valid restrictions with respect to the strong ordering. Hence, the output is U_{0j}^{*}, as defined in Section 2.1, with respect to this choice of ordering.

Lemma 2. *Given strongly ordered $P_1, P_2, \ldots, P_k$, we can solve ordered containment in time which is, as a function of k, $O(k)$.*

Proof. Let FIND_U_{0j}^{*} be a procedure consisting of the three loops in the proof of Lemma 1. Consider the following algorithm for a given strong ordering. Call FIND_U_{0j}^{*}. By Lemma 1, it creates U_{0j}^{*}, $1 \leq j \leq k$. If any $U_{0j}^{*} = \emptyset$, the algorithm reports infeasibility and terminates. Otherwise, select any j and any $t_j \in U_{0j}^{*}$. One can select positions t_i, $1 \leq i \neq j \leq k$, consistent with t_j, as follows:

$$\textbf{For } i \leftarrow j+1 \textbf{ to } k, \textbf{ choose } t_i \in U_{0i}^{*} \cap (U_{i-1,i}^{+}(C) + t_{i-1}).$$
$$\textbf{For } i \leftarrow j-1 \textbf{ to } 1, \textbf{ choose } t_i \in U_{0i}^{*} \cap (U_{i+1,i}^{-}(C) + t_{i+1}).$$

Lemma 1 shows that the cost of FIND_U_{0j}^{*}, as a function of k, is $O(k)$, so the total cost of the algorithm as a function of k is also $O(k)$. To establish that it yields a valid configuration, we first observe that polygons adjacent in the ordering are prevented from overlapping by the two operations $U_{i-1,i}^{+}(C) + t_{i-1}$ and $U_{i+1,i}^{-}(C) + t_{i+1}$. Because adjacent polygons cannot overlap and the ordering is strong, nonadjacent polygons cannot overlap. Thus, any configuration produced is valid. Now we need only show that the loops run to completion (i.e. $U_{0i}^* \cap (U_{i-1,i}^+(C) + t_{i-1}) \neq \emptyset$ and $U_{0i}^* \cap (U_{i+1,i}^-(C) + t_{i+1}) \neq \emptyset$). W.l.o.g. we discuss the first loop. The second loop of FIND_U_{0j}^* guarantees that $U_{0i} \subseteq (U_{0,i-1} \oplus U_{i-1,i}^+(C))$. By assumption, $U_{0,i-1}^*$, U_{0i}^*, and $U_{i-1,i}^+(C)$ are all nonempty, so we have: $U_{0i} \cap (U_{0,i-1} \oplus U_{i-1,i}^+(C)) \neq \emptyset$. Now, by the definition of $U_{0,i-1}^*$, for any $t_{i-1} \in U_{0,i-1}^*$, there exists $t_i \in U_{0,i}^*$ such that t_{i-1} and t_i are part of a valid configuration. We therefore have $t_i - t_{i-1} \in U_{i-1,i}^+(C)$. As $t_{i-1} + t_i - t_{i-1} = t_i$, we obtain $t_i \in U_{0,i-1} \oplus U_{i-1,i}^+(C)$, which establishes that $U_{0i}^* \cap (U_{i-1,i}^+(C) + t_{i-1}) \neq \emptyset$.

4 Compliant Containment

In a *compliant containment* problem, there are l polygons $P_1, P_2, \ldots, P_l$ placed in a container C and there are k additional polygons $P_{l+1}, P_{l+2}, \ldots, P_{l+k}$ which we also wish to place into the container. The polygons in the container are permitted to move but not "jump" over each other. The new polygons can be placed anywhere. This section gives a theoretical algorithm and a practical algorithm for compliant containment.

4.1 Compliant Containment Algorithm

The compliant containment algorithm sets up a containment problem for $P_1, P_2, \ldots P_{l+k}$. However, in the initial hypothesis, it replaces U_{ij}^{init} by $U_{ij}^{\text{compliant}} \subseteq U_{ij}^{\text{init}}$, $0 \leq i < j \leq l$. This restricts the motion of polygons $P_1, P_2, \ldots, P_l$. Then it runs the (ordinary) containment algorithm on this modified hypothesis.

If $P_1, P_2, \ldots, P_l$ are placed by the ordered containment algorithm of the previous section, then Lemma 1 shows how to obtain U_{0j}^*, $1 \leq j \leq l$, with respect to a given strong ordering. In this case, we can set $U_{ij}^{\text{compliant}}$, $1 \leq i < j \leq l$, equal to the following subset of U_{ij}^{init}, (which is a superset of U_{ij}^*): $U_{ij}^{\text{init}} \cap (U_{i0}^* \oplus U_{0j}^*)$.

In general, the containment algorithm (Section 2.2) can generate a convex cover of the set of valid placements of $P_1, P_2, \ldots, P_l$ into C. Each element of the cover is a hypothesis $\langle U_{ij}, 0 \leq i < j \leq l \rangle$ with all U_{ij} convex. The current placement $\langle t_1^0, t_2^0, \ldots, t_l^0 \rangle$ of $P_1, P_2, \ldots, P_l$ (the input to compliant containment) belongs to exactly one of these ($t_j^0 - t_i^0 \in U_{ij}$, $0 \leq i < j \leq l$), but others may cover the same connected component of valid solutions (the set of solutions reachable by a nonoverlapping motion of the polygons). For each of these hypotheses, the compliant containment algorithm substitutes the convex U_{ij} in place of U_{ij}^{init}, $0 \leq i < j \leq l$, and then solves containment for $P_1, P_2, \ldots, P_{l+k}$.

4.2 Practical Compliant Containment

The general compliant containment algorithm must solve many restricted containment problems. In practice, we consider only a single list $\langle U_{ij}^{\text{compliant}}, 0 \leq i < j \leq l \rangle$, where $U_{ij}^{\text{compliant}}$ is a superset of the union of the convex cover. This list may correspond to more than one connected component in the space of valid solutions. However, for our purposes, there is no harm in generating excess solutions to containment (except for the extra time we use to generate them). Define $U^*_{ij}(1 \ldots l)$ to be the range of values of $t_j - t_i$ for all valid placements of $P_1, P_2, \ldots, P_l$ into C. For $0 \leq i < j \leq l$, we set $U^{\text{compliant}}_{ij}$ equal to the connected component of $U^*_{ij}(1 \ldots l)$ which contains $t^0_j - t^0_i$.

To truly compute $U^*_{ij}(1 \ldots l)$, we would again have to compute a convex cover of the valid placements of $P_1, P_2, \ldots, P_l$ in C. However, we can create a superset of $U^*_{ij}(1 \ldots l)$ as follows. Subdivide and restrict hypotheses for this containment problem as usual (no evaluation), but put a limit on the depth of the subdivision tree. Take the union of each U_{ij} from each "leaf" (undivided) hypothesis. We refer to this method as *subdivision restriction* (of the initial hypothesis) [14]. Subdivision restriction is a generalization of union restriction [1].

5 Oriented Strip Packing Heuristic

This section describes a column-based heuristic for oriented strip packing. Like our translational strip packing heuristic, our oriented strip packing heuristic is column-greedy. To select a length (x-dimension) for the column, it sorts the unplaced pieces in order of decreasing maximum dimension (height or length). The length of the chosen orientation of the first unplaced piece in the list determines the column length. The heuristic packs "long" pieces into the column using ordered containment, and it packs "short" pieces into the column using compliant containment. After each column is placed, it is helpful to apply leftward gravity, which is a form of compaction [11, 18, 12], to reduce fragmentation of the available space.[7]

5.1 Packing "Long" Polygons

The heuristic maintains a list of *choices*. Each *choice* has 1) a set of "long" polygons, 2) an orientation for each of these polygons, 3) a strong ordering on this set. The initial list consists of the (1, 2, 4, or 8) ways of orienting the first polygon. The width of the column for each choice is equal to the x-dimension of that orientation of the first polygon. The heuristic then creates a list of 2-polygon choices, then a list of 3-polygon choices, and so forth, until it cannot add another polygon (or none of the remaining polygons are "long"). For each list, the set of polygons in each choice is the same.

Suppose the heuristic has already created the list of 3-polygon choices. It creates the list of 4-polygon choices as follows. Let P be the next unplaced polygon. For each choice and for each legal orientation of P and for each of the (four) possible ways P can be inserted into the ordering of the three polygons

---

[7] Using repeated solutions to linear programs, it is possible to plan a nonoverlapping motion of a set of polygon that minimizes any linear potential function on the positions of the polygons. This process is called *compaction* [11, 18, 12]. *Leftward gravity* minimizes the sum of the x-coordinates of polygon positions.

of the choice, the heuristic determines if P can be inserted there using ordered containment. Out of all the 4-polygon choices it generates in this manner, it keeps the best N for each different column length, where we measure goodness by the amount of vertical freedom: the height of the U_{01}^* region, where P_1 is the first polygon in the ordering with respect to C. If no 4-polygon choices are generated, we try the next unplaced polygon.

5.2 Packing "Short" Polygons

After the heuristic has packed as many "long" polygons as possible, each of the "long" polygons is allowed only compliant motion within its corresponding U_{0i}^*. The heuristic continues to generate lists of choices by adding one piece at a time. Let P_i be the next unplaced piece. For each choice in the list of choices and for each legal orientation of P_i, the heuristic restricts (Section 2.2) the initial hypothesis of the compliant containment problem. The connected components of U_{0i} are sorted in order of increasing area. For each component of the restricted U_{0i}, in order, the algorithm solves the compliant containment problem with P_i restricted to that component until either P_i fits or all the components of U_{0i} have been examined. The new list of choices is the N best successful outcomes. If there are no successful solutions to compliant containment, the heuristic skips to the next unplaced piece. After P_i is added, it is restricted to compliant motion. Along with each new choice, we store the current valid position and set $U_{0i}^{\text{compliant}}$ of valid positions of P_i. We store U_{0i}^* for each "long" P_i and the selected component of U_{0i} for each "short" P_i.

6   Implementation and Results

We have implemented our translational strip packing heuristic. As mentioned in Section 1.1, this heuristic achieved a cloth utilization of 83.94% for a layout of 50 polygons given to us by a large automotive manufacturer. The cloth utilization achieved by the best automatic layout system in this manufacturer's competition was 85%. The dataset for this example is proprietary and is therefore not shown in this paper.[8]

Figure 2a shows a smaller example of the translational heuristic applied to a 26-polygon layout from a large apparel manufacturer. This example was generated in less than one hour on a 50 MHz SPARCstation.[9] The cloth utilization of this example is 80.04% and leftward compaction does not improve it. Each polygon has up to 8 possible orientations. We supplied the translational heuristic with the same polygon orientations chosen by a human expert. The production-quality layout produced by a human expert has cloth utilization of 81.40%. Leftward compaction improves this to 81.41% (see Figure 2b).

---

[8] Information on how to contact the automotive manufacturer will be supplied by the authors upon request.

[9] SPARCstation is a trademark of SPARC, International, Inc., licensed exclusively to Sun Microsystems, Inc.

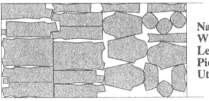

Name: 24409-3
Width: 57.00 in
Length: 114.93 in
Pieces: 26
Utilization: 80.04%

(a) Translational heuristic

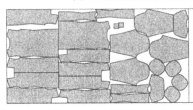

Name: 24409-3
Width: 57.00 in
Length: 113.00 in
Pieces: 26
Utilization: 81.41%

(b) Human expert

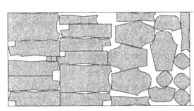

Name: 24409-3
Width: 57.00 in
Length: 113.00 in
Pieces: 26
Utilization: 81.41%

(c) Oriented heuristic

Fig. 2. Compacted strip packings

We have implemented ordered containment and "practical" compliant containment, as well as our oriented strip packing heuristic. We currently retain the best 5 choices of orientation and strong ordering for a set of "long" polygons. It is cheaper to run the ordered containment algorithm without checking strongness and then verify the output, which is what we implemented. For any particular output configuration $\langle t_1, t_2, \ldots, t_k \rangle$, it takes $O(k \log k)$ time to verify that $P_i + t_i$, $1 \leq i \leq k$, is a solution to containment.

Our current implementation of compliant containment uses only the simplest version of subdivision-restriction to determine the valid region for P_i (only restriction, no subdivision). Within compliant containment we currently limit the number of hypotheses examined to 10. (Note that a *hypothesis*, as defined in Section 2.2, is quite different from the notion of a *choice*). This reduces the running time but it occasionally prevents the algorithm from finding a position for a piece within a column. To compensate for this, after all columns have been placed we check the sparseness of the last column. If that column is too loose, we remove the column and attempt to place its pieces in previous columns using our *shrink-fit* technique.[10] This technique forms a scaled-down version of a piece and finds the nonoverlapping positions for it within the layout. It then uses overlap reduction to check if the unscaled piece fits at one of those locations.

For the 26-polygon example of Figure 2b, the layout achieved using our new algorithm is shown in Figure 2c. Note that the cloth utilization after compaction is the same as that of the compacted human-generated layout. Our result was produced in approximately 4 hours on a 50 MHz SPARCstation. No shrink-fit postprocessing was used. It might appear that the 81.41% cloth utilization obtained by both the human and our new algorithm is optimal for this dataset. One way to test this is to remove part of the end of the layout, guess a shorter length, and then run our ordinary translational containment algorithm using the shorter container. (In fact, if k is small, one can find the optimal strip length for a translational strip packing problem via binary search, as Daniels showed in her Ph.D. thesis [1].) For our 26-polygon example, we removed the last two columns (13 pieces) of the layout generated by our translational heuristic (Figure 2a) and then generated a "super-human" layout by starting with the length of the compacted human-generated layout as a goal, successfully solving containment, and then compacting. The result has cloth utilization of 81.60% and is shown in Figure 3.

Name: 24409-3
Width: 57.00 in
Length: 112.74 in
Pieces: 26
Utilization: 81.60%

Fig. 3. Compacted result of translational heuristic followed by translational containment on pieces from last 2 columns

KGR, a manufacturer of women's apparel in Lawrence, Massachusetts, supplied us with two sets of production markers (layouts) created by human experts with at least ten years of experience. Table 1 and Table 2 list cloth utilization from the first and second test sets, respectively. The running time of our algorithm on these examples is approximately one hour per column, which is roughly five times slower than the human.[11]

For the first set of markers, our algorithm matched or exceeded human cloth utilization on 11 of 19 examples (without shrink-fit post-processing for a sparse last column, we achieved this for only 5 of 19). Figure 4 shows one such "super-human" example for 23 items. For markers in the first set, the KGR marker makers sometimes constrain the legal orientations of the polygons beyond what

---

[10] The *shrink-fit* technique is part of our Automatic Marker Making Toolkit.

[11] With increases in the speed of the underlying polygon operations, we expect that the running time of our prototype software will improve in the near future.

Table 1. First KGR test suite: cloth utilization (in %) of compacted layouts produced by expert KGR marker makers vs. ordered and compliant containment heuristic vs. heuristic followed by shrink-fit post-processing

Example	k	KGR	Heuristic	Heuristic+
54303-dn	8	75.99	76.32	
54303-cn	8	71.60	73.59	
54303-8q	8	79.30	79.30	
54303-en	16	84.71	84.21	
54303-pn	16	83.95	84.14	
72631-4q	16	81.90	77.75	77.75
72631-3m	16	81.09	77.70	
72631-tn	16	81.76	81.24	
14349-3j	18	84.85	79.96	
14349-fi	18	85.18	75.22	83.41
14349-wm	18	81.14	65.28	
34045-xq	18	73.66	73.50	75.61
34045-ym	18	73.84	73.41	75.49
34045-zu	18	73.98	71.29	74.51
74375-fn	23	78.45	77.44	77.87
74375-ru	23	79.33	75.96	80.06
54303-5m	24	76.80	76.89	
54303-6q	24	82.54	79.97	82.85
54303-7u	24	85.66	79.40	85.68

Table 2. Second KGR test suite: cloth utilization (in %) of compacted layouts produced by expert KGR marker makers vs. ordered and compliant containment heuristic followed by shrink-fit post-processing

Example	k	KGR	Heuristic+
797748MF1A7C	10	72.33	71.85
74373MF1A3C	10	67.87	67.87
74373ML1A7A	14	78.25	72.16
74373ML1A3A	14	76.14	77.85
74373MF1A7B	20	76.37	68.86
74373MS1A3C	22	73.33	75.41
74373MS1A7C	22	76.80	78.11
799948ML1A3B	30	84.12	79.93

is given in the data files in order to increase the quality of a garment. Without these additional constraints, it is possible that they might be able to improve their cloth utilization for these markers. For markers in the second set, the marker makers used only the constraints in the data files. These markers are ones which were particularly challenging for the humans. Our algorithm met or exceeded human performance on 4 of 8 examples in this set. Two examples are shown in Figure 5 and Figure 6. Figure 7 shows an example in which the human expert produces the superior layout.

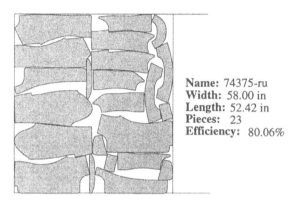

Name: 74375-ru
Width: 58.00 in
Length: 52.42 in
Pieces: 23
Efficiency: 80.06%

Fig. 4. Compacted result of oriented heuristic followed by shrink-fit post-processing for example 74375-ru of first KGR test suite

7 Future Work

Although our oriented strip packing software is still in an experimental phase, we view our new heuristic as a promising way to construct column-based layouts. In the future, we will consider adding multiple "short" polygons at the same time. Another possible improvement would allow the heuristic to "look-ahead" and assess the impact of current column choices on the packing of future columns. Our column-based pants panel layout algorithm [17] has this feature. Its look-ahead process tries to satisfy a "jaggedness" criterion for the end of the final column. We also plan to experiment with ways to reduce the column-greediness of the heuristic. One way is to "backtrack" (i.e. undo one or more columns). This feature exists in our panel layout algorithm. Alternatively, we can maintain multiple choices for each column as we place additional columns.

This work supports our thesis that good containment algorithms can lead to practical layout heuristics. It also emphasizes the value of interacting with industry. The layout from the auto industry motivated us to develop our first column-based heuristic. The data from KGR led us to develop the ordered containment and compliant containment algorithms and the second heuristic. Hence,

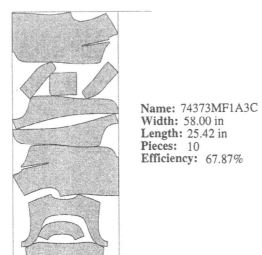

Name: 74373MF1A3C
Width: 58.00 in
Length: 25.42 in
Pieces: 10
Efficiency: 67.87%

Fig. 5. Compacted result of oriented heuristic for example 74373MF1A3C of second KGR test suite

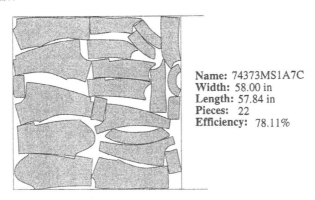

Name: 74373MS1A7C
Width: 58.00 in
Length: 57.84 in
Pieces: 22
Efficiency: 78.11%

Fig. 6. Compacted result of oriented heuristic for example 74373MS1A7C of second KGR test suite

our interaction with industry yielded new theory, algorithms, and heuristics. We plan to continue such fruitful interactions in the future.

References

1. K. Daniels. *Containment Algorithms for Nonconvex Polygons with Applications to Layout.* PhD thesis, Harvard University, 1995.
2. K. Daniels and V. J. Milenkovic. Multiple Translational Containment, Part I: An Approximate Algorithm. *Algorithmica, special issue on Computational Geometry in Manufacturing, to appear.*

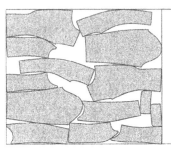

Name: 74373ML1A7A
Width: 44.00 in
Length: 52.04 in
Pieces: 14
Efficiency: 78.26%

(a) Human expert

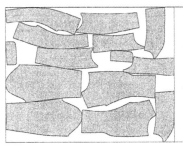

Name: 74373ML1A7A
Width: 44.00 in
Length: 56.43 in
Pieces: 14
Efficiency: 72.17%

(b) Oriented heuristic

Fig. 7. Compacted results of human expert vs. oriented heuristic followed by shrink-fit post-processing for example 74373ML1A7A of second KGR test suite

3. K. Daniels and V. J. Milenkovic. Multiple Translational Containment: Approximate and Exact Algorithms. In *Proceedings of the 6th Annual ACM-SIAM Symposium on Discrete Algorithms*, pages 205–214, 1995.

4. K. Daniels, V. J. Milenkovic, and Z. Li. Multiple Containment Methods. Technical Report 12–94, Center for Research in Computing Technology, Division of Applied Sciences, Harvard University, 1994.

5. K. A. Dowsland and W. B. Dowsland. Packing Problems. *European Journal of Operational Research*, 56:2 – 14, 1992.

6. H. Dyckhoff. A typology of cutting and packing problems. *European Journal of Operations Research*, 44:145–159, 1990.

7. L. Guibas, L. Ramshaw, and J. Stolfi. A Kinetic Framework for Computational Geometry. In *Proceedings of the 24th IEEE Symposium on Foundations of Computer Science*, pages 100–111, 1983.

8. O. Gurel. Additional considerations on marker layout problem by graph theory. Technical Report 320-2945, IBM Scientific Centre, 1968.

9. O. Gurel. Marker layout via graph theory: An attempt for optimal layout of irregular patterns. Technical Report 320-2921, IBM Scientific Centre, 1968.

10. O. Gurel. Circular graph of marker layout. Technical Report 320-2965, IBM Scientific Centre, 1969.

11. Z. Li. *Compaction Algorithms for Non-Convex Polygons and Their Applications.* PhD thesis, Harvard University, Division of Applied Sciences, 1994.
12. Z. Li and V. J. Milenkovic. A Compaction Algorithm for Non-Convex Polygons and Its Application. In *Proceedings of the 9th Annual ACM Symposium on Computational Geometry*, pages 153–162, May 1993.
13. V. J. Milenkovic. Multiple Translational Containment, Part II: Exact Algorithms. *Algorithmica, special issue on Computational Geometry in Manufacturing, to appear.*
14. V. J. Milenkovic. A Symmetry Breaking Restriction. Personal communication, March 1995.
15. V. J. Milenkovic. Translational Polygon Containment and Minimal Enclosure using Linear Programming Based Restriction. In *Proceedings of the 28th Annual ACM Symposium on the Theory of Computing*, 1996.
16. V. J. Milenkovic and K. Daniels. Translational Polygon Containment and Minimal Enclosure using Geometric Algorithms and Mathematical Programming. Technical Report 25-95, Center for Research in Computing Technology, Division of Applied Sciences, Harvard University, 1995.
17. V. J. Milenkovic, K. Daniels, and Z. Li. Placement and Compaction of Nonconvex Polygons for Clothing Manufacture. In *Proceedings of the 4th Canadian Conference on Computational Geometry*, pages 236–243, 1992.
18. V. J. Milenkovic and Z. Li. A Compaction Algorithm for Nonconvex Polygons and Its Application. *European Journal of Operations Research*, 84:539–560, 1995.
19. H. Minkowski. Volumen und Oberfläche. *Mathematische Annalen*, 57:447–495, 1903.
20. J. Serra. *Image Analysis and Mathematical Morphology*, volume 1. Academic Press, New York, 1982.
21. J. Serra, editor. *Image Analysis and Mathematical Morphology*, volume 2: Theoretical Advances. Academic Press, New York, 1988.
22. P. E. Sweeney and E. R. Paternoster. Cutting and Packing Problems: A Categorized, Application-Oriented Research Bibliography. *Journal of the Operational Research Society*, 43(7):691–706, 1992.

Computing a Flattest, Undercut-Free Parting Line for a convex Polyhedron, with Application to Mold Design *

Jayanth Majhi[1]   Prosenjit Gupta[2]   Ravi Janardan[3]

[1] Dept. of Computer Science, Univ. of Minnesota, Minneapolis, MN 55455;
majhi@cs.umn.edu
[2] Max-Planck-Institut für Informatik, Im Stadtwald, D-66123 Saarbrücken, Germany.
pgupta@mpi-sb.mpg.de.
[3] Dept. of Computer Science, Univ. of Minnesota, Minneapolis, MN 55455;
janardan@cs.umn.edu

Abstract. A parting line for a convex polyhedron, $\mathcal{P}$, is a closed curve on the surface of $\mathcal{P}$. It defines the two pieces of $\mathcal{P}$ for which mold-halves must be made. An undercut-free parting line is one which does not create recesses or projections in $\mathcal{P}$ and thus allows easy de-molding of $\mathcal{P}$. Computing an undercut-free parting line that is as flat as possible is an important problem in mold design. In this paper, an $O(n^2)$-time algorithm is presented to compute such a line, according to a prescribed flatness criterion, where n is the number of vertices in $\mathcal{P}$.

1   Introduction

We consider a geometric problem which arises in the design of molds for casting and injection molding—two ubiquitous manufacturing processes. Consider, for instance, the construction of a sand mold for casting a solid. First a prototype, $\mathcal{P}$, of the solid is made. Two halves of $\mathcal{P}$ are then identified and a separate mold-half is made for each. This involves placing $\mathcal{P}$ in a mold-box and packing sand around the first half. After the sand has been compacted and hardened, $\mathcal{P}$ is translated out of the mold-half (i.e., de-molded) and a second mold-half is made similarly for the other half of $\mathcal{P}$. The two mold-halves are then fastened together by pins to form a cavity in the shape of $\mathcal{P}$ and molten metal is poured into it. (More details can be found in [15].)

Among the many key issues surrounding the design of a good mold, two that are mentioned extensively in the literature are (i) the shape of the parting line and (ii) the number of undercuts.

The *parting line* is a continuous closed curve on the surface of $\mathcal{P}$ which defines the two halves; thus it also defines the profile of the contact surface between the two mold-halves. As noted in [1, 15], the parting line should be chosen to be as "flat" as possible since this results in a more cost-efficient and accurate mold.

---

* Extended abstract. Research supported in part by NSF Grant CCR–9200270 and by a Univ. of Minnesota Grant–in–Aid of Research Award.

(Loosely speaking, by a "flat" parting line, we mean one which lies as nearly as possible in a plane. We will formalize this notion in Section 1.1.) An *undercut* is a recess or projection in $\mathcal{P}$ which prevents its easy de-molding. This can be alleviated by using cores and inserts. However, this is generally discouraged since it increases the cost of the mold and slows down part production [6, 15]. Thus it is important to minimize the number of undercuts. This can be accomplished by choosing the two halves of $\mathcal{P}$ carefully.

Not surprisingly, these two issues are often mutually conflicting, even if $\mathcal{P}$ is convex: For the octahedron in Figure 1, any plane which divides it into two creates undercuts—e.g., the plane containing vertices 1, 2, and 3 creates an undercut for the upper half (a projection under the chain of vertices 1–4–3). Undercuts can be avoided by choosing the parting line to be the chain 1–2–3–4–1 (or 2–5–4–6–2); however, the parting line is no longer flat.

1.1 Formalizing the problem

In this paper, we consider the problem of finding the flattest, undercut–free parting line for a convex polyhedron $\mathcal{P}$. We now formalize our notion of "parting line" and "flatness".

Suppose that $\mathcal{P}$ is viewed from infinity along a direction $\mathbf{d}$. A point p on $\mathcal{P}$ is $\mathbf{d}$-*visible* if the ray from p in direction $-\mathbf{d}$ misses the interior of $\mathcal{P}$; a facet f of $\mathcal{P}$ is $\mathbf{d}$-*visible* if every point of f is $\mathbf{d}$-visible. Note that if a facet f is $\mathbf{d}$-visible, then it will not create an undercut when $\mathcal{P}$ is de-molded along directions $\mathbf{d}$ (and $-\mathbf{d}$), since no part of $\mathcal{P}$ obstructs f in direction $\mathbf{d}$.

Let $L(\mathbf{d})$ be a parting line for $\mathcal{P}$ w.r.t. de-molding along directions $\mathbf{d}$ and $-\mathbf{d}$. $L(\mathbf{d})$ consists of line segments, which are defined as follows:

Suppose that $\mathcal{P}$ has no facets parallel to $\mathbf{d}$ (we call these $\mathbf{d}$-*parallel facets*). Let $F(\mathbf{d})$ be the set of $\mathbf{d}$-visible facets and let $B(\mathbf{d})$ be the boundary of their union. $B(\mathbf{d})$ is a closed chain of edges of $\mathcal{P}$. In this case, each line segment of $L(\mathbf{d})$ is an edge of $B(\mathbf{d})$.

Suppose $\mathcal{P}$ has one or more $\mathbf{d}$-parallel facets, f. By definition, $f \in F(\mathbf{d})$. Now $B(\mathbf{d})$ is of the form $c_1 F_1 c_2 F_2 \ldots c_m F_m$, for some m. Here each c_i is a chain of edges of $\mathcal{P}$, with endpoints u_i and v_i and F_i is a group of contiguous $\mathbf{d}$-parallel facets, attached to c_i at v_i and to c_{i+1} at u_{i+1}—indices taken modulo m. (See Figure 2.) Let p_i be a Euclidean shortest path lying in F_i and joining v_i and u_{i+1}. In this case, each line segment of $L(\mathbf{d})$ is an edge of a c_i or an edge of a p_i, $1 \leq i \leq m$.

In general then, $L(\mathbf{d}) = e_1 e_2 \ldots e_k$, for some k, where each e_i is a line segment as defined above. It is easy to see that $L(\mathbf{d})$ does not create undercuts. Let $\hat{L}(\mathbf{d}) = \hat{e}_1, \hat{e}_2, \ldots, \hat{e}_k$ be the projection of $L(\mathbf{d})$, along direction $\mathbf{d}$, onto a plane normal to $\mathbf{d}$. Note that while $\hat{L}(\mathbf{d})$ lies in a plane, $L(\mathbf{d})$ can be highly stepped since the e_i can zigzag considerably in 3-space. Our measure of the flatness of $L(\mathbf{d})$ is:

$$\rho(\mathbf{d}) = \sum_{i=1}^{k} length(\hat{e}_i)^2 \Big/ \sum_{i=1}^{k} length(e_i)^2, \qquad (1)$$

where *length*(l) is the Euclidean length of segment l. Note that $\rho(\mathbf{d}) \leq 1$, with equality holding if and only if $L(\mathbf{d})$ lies in a plane.[4] In general, the larger the value of $\rho(\mathbf{d})$, the flatter is $L(\mathbf{d})$. We can now state our problem formally:

Problem 1. Given a convex polyhedron $\mathcal{P}$ with n vertices, find a direction $\mathbf{d}$ such that $\rho(\mathbf{d})$ is maximized.

1.2 Overview of the result and prior related work

We give an algorithm for Problem 1 which runs in $O(n^2)$ time. Our algorithm is an interesting blend of discrete and continuous optimization. Briefly, our approach is as follows: We first subdivide 3-space into $O(n^2)$ unbounded polyhedral regions (called *cones*), each apexed at the origin. Each cone has the property that $L(\mathbf{d})$ is the same for all points $\mathbf{d}$ in the cone's interior. Thus, maximizing $\rho(\mathbf{d})$ inside a cone is equivalent to maximizing the numerator in Equation 1, which gives rise to a continuous optimization problem for each cone. (The optimization problem is discussed in detail in Section 2.2.)

Similarly, $L(\mathbf{d})$ is the same for all directions $\mathbf{d}$ that lie on a bounding plane of a cone. However, now $\mathcal{P}$ will have $\mathbf{d}$-parallel facets, so that we need to first perform certain shortest path computations on the surface of $\mathcal{P}$ to compute $L(\mathbf{d})$ before proceeding to formulate and solve an optimization problem as above. It turns out that the former problem can be formulated as a shortest path problem on a special planar polygon and hence can be solved quickly.

Thus, the idea is to compute the ρ-value for each cone and cone boundary and pick the best one. However, a direct implementation of this method is not efficient because formulating an optimization problem for a cone or cone boundary requires knowledge of the parting line, and computing the latter from scratch each time takes $O(n)$ per parting line and results in an $O(n^3)$ algorithm. We circumvent this problem by visiting the cones in a certain order and updating the parting line incrementally, so that the total time is $O(n^2)$.

We discuss the algorithm in more detail in Section 2. We have also implemented a version of this algorithm; we discuss this implementation and our computational experience with it in Section 3.

1.3 Related work

We are not aware of prior work on the specific problem that we consider; however, we mention briefly some related work: Bose, Bremner, and van Kreveld [2] (see also [3]) present efficient algorithms to decide if a given polyhedron admits an undercut–free parting line which lies in a plane. For a convex (resp. non-convex) polyhedron, their algorithm runs in time $O(n \log^2 n)$ (resp. $O(n^{3/2+\epsilon})$), where $\epsilon > 0$ is an arbitrarily small constant. In [6], Chen, Chou, and Woo consider the problem of finding a de-molding direction which minimizes the number of

---

[4] Our choice of sum of squared lengths instead of sum of lengths is for technical reasons and is discussed further in Section 3.

undercuts. For an n-vertex polyhedron, P, their algorithm takes time $O(np \log p)$, where p is the number of pockets in P. (A *pocket* of P is a connected region of $CH(P) - P$, where $CH(P)$ is the convex hull of P.) However, they do not consider the shape of the parting line, which can be quite stepped. In [10], Hui and Tan give heuristics for finding parting directions with few (but not necessarily minimum) undercuts. Ravi and Srinivasan [14] identify several criteria (different from ours) for the design of good parting lines but do not give any algorithms for computing these lines. In [13], related decomposition problems are addressed for 2-dimensional molds.

We close by mentioning some other geometric work of interest in the area of mold design. In [3, 4, 5], the general problem of "mold fillability" is addressed. The questions of interest here include deciding whether a given mold can be filled from a given "pour" direction without creating air pockets, determining all such pour directions, computing a direction that minimizes the number of air pockets (if air pockets are unavoidable), and characterizing classes of polyhedra with respect to their fillability. Efficient algorithms are given in [3, 4] for 2-dimensional molds and in [3, 5] for 3-dimensional molds. In a working paper [8], related questions are also addressed for different mold-filling strategies and different filling materials.

2 The algorithm

2.1 Subdividing 3-space into cones

We follow the approach in [12]: For each facet f of $\mathcal{P}$, we construct a plane, h_f, which is parallel to f and passes through the origin. The planes h_f subdivide 3-space into the afore-mentioned collection of cones. (See Figure 3.) Let $\mathbf{S}^2$ be the unit-sphere centered at the origin. Let g_f be the great circle $h_f \cap \mathbf{S}^2$ and let $\mathcal{A}$ be the arrangement of the g_f's on $\mathbf{S}^2$. $\mathcal{A}$ has $O(n^2)$ vertices, edges, and faces, and it is easy to establish a 1–1 correspondence between the unbounded cones and the faces of $\mathcal{A}$. Therefore, we will hereafter work with $\mathcal{A}$; for convenience, however, we will continue to refer to the faces of $\mathcal{A}$ as cones. Note that a direction d is now a point on $\mathbf{S}^2$ and hence a unit-vector.

Lemma 2. $L(\mathbf{d})$ *is the same for all directions* $\mathbf{d}$ *in the interior of a cone* C. *Similarly, if* e *is an edge of* C, *then* $L(\mathbf{d})$ *is the same for all directions* $\mathbf{d}$ *in the interior of* e.

We denote the unique parting line associated with the interior of C (resp. interior of e) by $L(C)$ (resp. $L(e)$). Trivially, if v is a vertex of C, then the associated parting line is unique; we denote this by $L(v)$.

2.2 The optimization problem

Let $d \in \mathbf{S}^2$ and let C be any cone. Let $L(C) = \mathbf{e_1}, \mathbf{e_2}, \ldots, \mathbf{e_k}$, where each $\mathbf{e_\ell}$ is an edge of $\mathcal{P}$ (with orientation assigned arbitrarily). Assume that $\mathbf{d} = x\mathbf{i} + y\mathbf{j} + z\mathbf{k}$

and $e_\ell = a_\ell \mathbf{i} + b_\ell \mathbf{j} + c_\ell \mathbf{k}$, where $\mathbf{i}$, $\mathbf{j}$, and $\mathbf{k}$ are unit-vectors along the x-, y-, and z-axes. Let θ_ℓ be the angle between e_ℓ and $\mathbf{d}$.

We have $length(\hat{e}_\ell)^2 = (length(e_\ell) \cdot \sin \theta_\ell)^2 = (length(e_\ell \times \mathbf{d}))^2$. Since $e_\ell \times \mathbf{d} = (b_\ell z - c_\ell y)\mathbf{i} + (c_\ell x - a_\ell z)\mathbf{j} + (a_\ell y - b_\ell x)\mathbf{k}$, we have $length(\hat{e}_\ell)^2 = A_\ell x^2 + B_\ell y^2 + C_\ell z^2 + D_\ell xy + E_\ell yz + F_\ell xz$, where $A_\ell = c_\ell^2 + b_\ell^2$, $B_\ell = a_\ell^2 + c_\ell^2$, $C_\ell = a_\ell^2 + b_\ell^2$, $D_\ell = -2a_\ell b_\ell$, $E_\ell = -2b_\ell c_\ell$, and $F_\ell = -2a_\ell c_\ell$. Thus, $\sum_{\ell=1}^{k} length(\hat{e}_\ell)^2 = Ax^2 + By^2 + Cz^2 + Dxy + Eyz + Fzx$, where $A = \sum_{\ell=1}^{k} A_\ell$, and similarly for B through F.

Let $d_C \in \mathbf{S}^2$ be a given point in C's interior. (d_C is computed at the time $\mathcal{A}$ is constructed.) Let n_f be the outward-directed normal to facet f. Note that d is in the interior of C if and only if $\mathbf{d} \cdot \mathbf{n}_f$ and $\mathbf{d}_C \cdot \mathbf{n}_f$ are both positive or both negative for each great circle g_f bounding C.

Thus our optimization problem for the interior of C is:

Maximize $\quad f(x, y, z) = Ax^2 + By^2 + Cz^2 + Dxy + Eyz + Fxz$

Subject to $\quad x^2 + y^2 + z^2 = 1$ $\quad$ (*Sphere Constraint.*)
$\quad\quad\quad\quad\quad$ $\mathbf{d} \cdot \mathbf{n}_f > 0$ (resp. < 0) if $\mathbf{d}_C \cdot \mathbf{n}_f > 0$ (resp. < 0) for each great
$\quad\quad\quad\quad\quad$ circle g_f bounding C. (*Plane Constraints.*)

The optimization problem for the interior of an edge e is formulated similarly. However, there are just three plane constraints now—one requiring that the solution point lie on e's great circle and the other two requiring that it lie in the interior of e. For a vertex v, there is no need to solve any optimization problem, since it is a single point.

2.3 Solving the optimization problem

We use the method of Lagrange multipliers [11]. In more detail, consider the optimization problem for a cone C. The Lagrangian is $L(x, y, z, \lambda) = f(x, y, z) + \lambda(1 - x^2 - y^2 - z^2)$, for some parameter λ. The partial derivatives of L, w.r.t. each of x, y, and z, must be zero at an extreme (i.e., minimum or maximum) point. This yields three linear equations in x, y, and z. The values of λ for which these three equations have non-trivial solutions can be found by solving a cubic equation in λ, given by:

$$\begin{vmatrix} 2A - 2\lambda & D & F \\ D & 2B - 2\lambda & E \\ F & E & 2C - 2\lambda \end{vmatrix} = 0.$$

For each such real-valued λ (there are at most three of them) we solve for x, y, and z, using any two of the three linear equations (the remaining one will depend on the two chosen) and the sphere constraint. This will yield (i) two antipodal points on $\mathbf{S}^2$, or (ii) a great circle (if the three equations are the same but not identically zero), or (iii) all of $\mathbf{S}^2$ (if the three equations are identically zero). We can ignore cases (ii) and (iii) since, anyway, we will later be computing the parting line at edge interiors and at vertices. If case (i) holds then we check if either of the two points lies in C (by checking the plane constraints) and, if so, then we compute the maximum value of ρ in the interior of C.

For the interior of an edge, e, which lies on a great circle defined by the plane $ax + by + cz = 0$, the Lagrangian is $L(x, y, z, \lambda_1, \lambda_2) = f(x, y, z) + \lambda_1(1 - x^2 - y^2 - z^2) + \lambda_2(ax + by + cz)$, for some parameters λ_1 and λ_2. Setting partial derivatives to zero gives three linear equations in x, y, z, and λ_2. Using these equations and the equation $ax + by + cz = 0$ we can compute the values of λ_1 that yield non-trivial solutions, this time by solving a quadratic equation in λ_1, as given by:

$$\begin{vmatrix} 2A - 2\lambda_1 & D & F & a \\ D & 2B - 2\lambda_1 & E & b \\ F & E & 2C - 2\lambda_1 & c \\ a & b & c & 0 \end{vmatrix} = 0.$$

We can now eliminate λ_2 using one of the linear equations. Using the sphere constraints, the constraint $ax + by + cz = 0$, and any one of the remaining linear equations, we proceed to compute the extreme points and find the maximum value of ρ in the interior of e.

Analysis: It is reasonable to assume that the cubic and quadratic equations that arise can be solved in $O(1)$ time. Thus, the optimization problem for C (resp. e) takes time $O(|C|)$ (resp. $O(1)$). Summed over all cones and edges, this is $O(n^2)$.

Remark. If some other optimization algorithm is to be used, it would be advantageous to first convert our optimization problem to one of constant size, i.e., one where all of the following are constant: (i) the number of variables, (ii) the description size of the objective function and of each constraint, and (iii) the number of constraints. It might then be reasonable to assume that each such problem can be solved in constant time, for a time bound of $O(n^2)$ for all problems. Conditions (i) and (ii) already hold in our case. To enforce condition (iii), we can triangulate the cones into a total of $O(n^2)$ subcones, which is easy to do since the cones are convex. The number of optimization problems is still $O(n^2)$, but now each has just four constraints.

2.4 Handling d-parallel facets

In the presence of **d**-parallel facets, portions of the parting line need to be computed as shortest paths (Figure 2). $\mathcal{P}$ has **d**-parallel facets if **d** is in the interior of an edge $e \in \mathcal{A}$ or if **d** is a vertex of $\mathcal{A}$. In the former case, there are at most two groups, F_1 and F_2, of **d**-parallel facets, and each F_i consists of just one facet f_i, $i = 1, 2$. Thus, p_i is just the line segment $\overline{v_i u_{i+1}}$.

In the latter case, there can be many groups F_i, each made up of several contiguous **d**-parallel facets. The computation of p_i can now be formulated as a 2-dimensional shortest path problem inside a simple polygon, as follows: Let $\overline{a_1 b_1}, \ldots, \overline{a_t b_t}$ ($t \geq 1$) be the edges of $\mathcal{P}$ shared by successive facets of F_i, where

$\overline{a_1 b_1}$ is closest to v_i and $\overline{a_t b_t}$ is closest to u_{i+1}. Note that these edges are necessarily vertical (w.r.t. **d**). Consider the portion H_i of F_i which is enclosed between the chains $v_i a_1 \ldots a_t u_{i+1}$ and $v_i b_1 \ldots b_t u_{i+1}$. (See Figure 4.) From the triangle inequality, it is clear that any shortest (v_i, u_{i+1})-path lying in F_i must also lie in H_i. We can "flatten" H_i on the plane without changing edge lengths to get a simple polygon, which we continue to call H_i. (In fact, H_i is monotone in the direction **d**.) Our problem now becomes one of finding a shortest (v_i, u_{i+1})-path in H_i, which we can do in $O(|H_i|)$ time using the algorithm in [9]. (This algorithm requires that H_i be triangulated in linear time. This can be done by simply adding the segments $\overline{a_i b_i}$ $(1 \le i \le t)$ and $\overline{a_i b_{i+1}}$ $(1 \le i \le t-1)$.)

Computing a shortest loop An important special case arises if all the facets contributing to the parting line are **d**-parallel. Let Q denote the circular sequence of these facets. Now, $L(\mathbf{d})$ is a shortest closed chain of line segments (i.e., a *loop*) lying in Q. (See Figure 5.) However, it is not clear how to compute it since we do not have, at this point, a start vertex and an end vertex between which to run the shortest path algorithm—in general the shortest loop could potentially cross each of the shared vertical edges anywhere.

Fortunately, Lemma 3 below establishes that there is always a shortest loop which passes through a certain vertex of Q. Let $\overline{a_1 b_1}, \ldots, \overline{a_t b_t}$ $(t \ge 1)$ be the vertical edges shared by successive facets of Q, where b_i is below a_i for all i. Let A (resp. B) be the closed chain of edges $\overline{a_1 a_2}, \ldots, \overline{a_t a_1}$ (resp. $\overline{b_1 b_2}, \ldots, \overline{b_t b_1}$). Let Q' be the portion of Q lying between A and B. By the triangle inequality, it follows that any shortest loop lying in Q must also lie in Q'.

Lemma 3. *There is a shortest loop around Q which passes through the lowest vertex, a_l, of A.*

Proof. Let J be any shortest loop around Q—hence J lies in Q'. We first claim that if J bends upwards (resp. downwards), then it can do so only at a vertex of A (resp. B). Why? Assume that J bends upwards at a point p. (The discussion is similar if J bends downwards.) For some i, p lies in the trapezoid of Q' defined by the vertices a_i, a_{i+1}, b_i, and b_{i+1}. If p is any point of this trapezoid other than a_i and a_{i+1}, then we can shorten J by picking two points on it that are sufficiently close to p and on opposite sides of the vertical line through p and joining them directly while still staying within Q' (hence within Q). This contradicts the optimality of J and establishes the claim.

For the rest of this proof we take J to be a highest possible shortest loop lying in Q', i.e., we take a shortest loop around Q' and push it upwards as far as possible without leaving Q'. By the above claim, it follows that J passes through at least one vertex of A. Let a_l be the lowest vertex of A and assume, for a contradiction, that J does not pass through a_l. Let $c \ne a_l$ be the point where J crosses $\overline{a_l b_l}$. At c, J has three possible orientations: It is (i) horizontal, (ii) sloping downwards to the right, or (iii) sloping upwards to the right.

Case (i) Since J must pass through a vertex of A and this vertex cannot be lower than a_l, it follows that as we walk along J to the right of c, J must bend upwards. Consider the first bend. By the claim above, this bend (or any other upward bend for that matter) can occur only at some vertex of A. But then this vertex is lower than a_l since J is horizontal at c and c is below a_l—a contradiction.

Case (ii) As we walk along J to the right of c, we move downwards. Since J is a loop it must bend upwards at some point in order to return to c. Consider the first such bend. By the above claim, this bend must be at a vertex of A. But then this vertex is lower than a_l—again a contradiction.

Case (iii) Similar to Case (ii), except that we walk along J to the left of c.

Lemma 3 gives us the desired start and end vertex for the shortest path algorithm. Specifically, we cut Q' along $\overline{a_l b_l}$ and flatten it out into a polygon in the plane; thus $\overline{a_l b_l}$ appears at the two ends of the polygon. We then run the algorithm of [9] between the two copies of a_l to find the shortest loop. This takes $O(|Q'|)$ time.

2.5 Incremental computation and overall analysis

We will only sketch the main ideas: We compute A by mapping the g_f's to the plane, via central projection, applying the algorithm of [7], and then inverting the projection, all in $O(n^2)$ time. We then proceed in two phases: In the first phase, we compute the $L(C)$'s for cones C and the $L(e)$'s for edges e, and in the second phase the $L(\mathbf{d})$'s, for vertices $\mathbf{d}$, again in $O(n^2)$ time. Clearly, all objective function updates can be done within this time bound too.

Let $\bar{A}$ be the (graph) dual of A. We do a depth-first search (dfs) of $\bar{A}$, starting at a vertex $\bar{C}$, corresponding to cone C. We can initialize $L(C)$ by computing all the visible facets and their boundary (i.e., $L(\bar{C})$) in $O(n)$ time. Let $\bar{C}'$ be the next vertex visited, via edge e'—i.e., cones C and C' share an edge e in A. As we move from C to a point $\mathbf{d}$ inside e, one of the two associated $\mathbf{d}$-parallel facets, say f_1, will come more fully into view, while the other, f_2, will start to disappear. We delete f_2's contribution to $L(C)$ and include $f \cap f_1$ for each facet f that is adjacent to f_1 and contributes to $L(C)$—this gives $L(e)$. Computing $L(C')$ from $L(e)$ is essentially the reverse. The total time can be shown to be $O(n \sum_{f \in \mathcal{P}} |f|) = O(n^2)$.

In the second phase, we do a dfs of A itself, starting at a vertex $\mathbf{d}$. It can be shown that $L(\mathbf{d})$ can be computed in $O(n)$ time, inclusive of the construction of the sequence $c_1 F_1 \ldots c_m F_m$, the H_i's, and the shortest paths p_i (Figures 2 and 4). Suppose that the dfs next visits an adjacent vertex $\mathbf{d}'$, along edge e. How does $L(\mathbf{d})$ change to $L(\mathbf{d}')$? As we "move along" e, some $\mathbf{d}$-parallel facets come more fully into view while others disappear, causing some of the F_i's to be replaced by chains. Also, some facets will become $\mathbf{d}'$-parallel, causing some c_j's to be replaced by groups of contiguous $\mathbf{d}'$-parallel facets. Thus, $c_1 F_1 \ldots c_m F_m$ is updated (somewhat analogous to phase 1) to get $c_1' F_1' \ldots c_s' F_s'$. Then the H_i's are constructed and the p_i's computed. Again, the total time can be shown to be $O(n^2)$. Combined with the analysis of the time used for solving the optimization problems (Section 2.3), this leads to our main result:

Theorem 4. *Problem 1 can be solved in $O(n^2)$ time.*

3   Implementation and discussion

We have implemented a version of the algorithm, which includes handling directions lying in cone and edge interiors and directions corresponding to vertices, but does not incorporate the incremental computation. Our goal was to compare the best and worst parting lines for a "typical" convex polyhedron. Figure 6 shows these (in heavy lines) for a 40-vertex polyhedron, where the best line has $\rho = 0.9452$ and the worst line has $\rho = 0.2980$. The polyhedron was generated using qhull (http://www.geom.umn.edu/software/download/qhull.html) to compute the convex hull of forty co-spherical points generated randomly. Our implementation is written in C++, runs on an SGI Irix 5 machine and uses LEDA (http://www.mpi-sb.mpg.de/LEDA) to compute $\mathcal{A}$ and the shortest paths.

Even without the incremental computation, the implementation runs very fast—for the 40-vertex polyhedron, it takes only 0.16 seconds to compute the optimal parting line (excluding the time needed for graphical output).

We close with a few remarks on our choice of ρ. In particular, the reason we chose squared edge lengths is that the resulting objective function has $O(1)$ terms, regardless of the size, k, of the parting line—k can be $\Theta(n)$ in the worst case. Had we used (unsquared) edge lengths, the objective function would have had k terms, each under a square-root. Thus, merely evaluating it at each cone interior, edge interior, and vertex would take $O(n)$ time and lead to an $O(n^3)$ algorithm. (Our implementation can be modified very easily to incorporate this version of the algorithm also.) Another remark concerns the choice of p_i as a Euclidean shortest path. Given our choice of squared lengths, a better choice for p_i would have been a sequence of edges, one per facet of F_i, such that the sum of the squares of the edge lengths is minimized. However, at this time, we do not know how to compute such a p_i efficiently; we are investigating this further.

References

1. R.W. Bainbridge. Thermo–setting plastic parts. In J.G. Bralla, editor, *Handbook of product design for manufacturing*, pages 6–3—6–16. McGraw–Hill Book Company, 1986.

2. P. Bose, D. Bremner, and M. van Kreveld. Determining the castability of simple polyhedra. In *Proceedings of the 10th Annual ACM Symp. on Computational Geometry*, pages 123–131, 1994.

3. P. Bose. *Geometric and computational aspects of manufacturing processes*. PhD thesis, School of Computer Science, McGill University, Montréal, Canada, 1995.

4. P. Bose and G. Toussaint. Geometric and computational aspects of gravity casting. *Computer–Aided Design*, 27(6):455–464, 1995.

5. P. Bose, M. van Kreveld, and G. Toussaint. Filling polyhedral molds. In *Proc. 3rd Workshop Algorithms Data Structures*, volume 709 of *Lecture Notes in Computer Science*, pages 210–221. Springer-Verlag, 1993.

6. L-L. Chen, S-Y. Chou, and T.C. Woo. Parting directions for mould and die design. *Computer-Aided Design*, 25(12):762–768, 1993.

7. H. Edelsbrunner, J. O'Rourke, and R. Seidel. Constructing arrangements of lines and hyperplanes, with applications. *SIAM Journal on Computing*, 15:341–363, 1986.

8. S. Fekete and J. Mitchell. Geometric aspects of injection molding. Manuscript, 1992.

9. L. Guibas, J. Hershberger, D. Leven, M. Sharir, and R.E. Tarjan. Linear-time algorithms for visibility and shortest path problems inside triangulated simple polygons. *Algorithmica*, 2:209–233, 1987.

10. K.C. Hui and S.T. Tan. Mould design with sweep operations–a heuristic search approach. *Computer-Aided Design*, 24(2):81–91, 1992.

11. D.G. Luenberger. *Introduction to linear and non-linear programming*. Addison-Wesley, 1973.

12. M. McKenna and R. Seidel. Finding the optimal shadows of a convex polytope. In *Proceedings of the 1st Annual ACM Symp. on Computational Geometry*, pages 24–28, 1985.

13. A. Rosenbloom and D. Rappaport. Moldable and castable polygons. In *Proceedings of the 4th Canadian Conference on Computational Geometry*, pages 322–327, 1992.

14. B. Ravi and M.N. Srinivasan. Decision criteria for computer-aided parting surface design. *Computer-Aided Design*, 22(1):11–18, 1990.

15. E.C. Zuppann. Castings made in sand molds. In J.G. Bralla, editor, *Handbook of product design for manufacturing*, pages 5–3—5–22. McGraw–Hill Book Company, 1986.

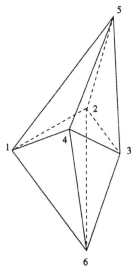

Figure 1: An octahedron which admits no undercut-free parting line lying in a plane

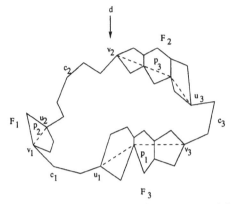

Figure 2: The parting line in the presence of d-parallel facets

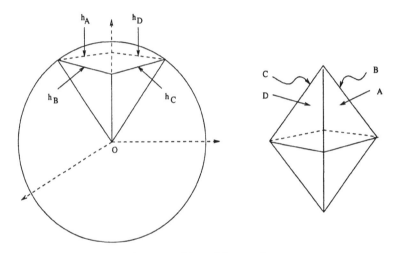

Figure 3: Decomposition of 3-space into cones

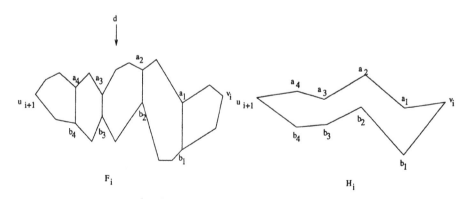

Figure 4: Formulating the 2-dimensional shortest path problem.

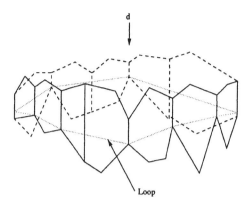

Figure 5: A loop around a circular sequence of d-parallel facets

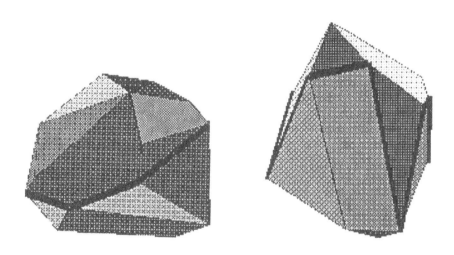

*Figure 6: Best and worst parting lines produced by the implementation for
an example 40-vertex polyhedron. The best line (left) has
rho=0.9452, while the worst line (right) has rho=0.2980*

Geometric Problems in Machine Learning

David Dobkin[1] and Dimitrios Gunopulos[2]

[1] Computer Science Dept., Princeton University,
35 Olden St., Princeton, NJ, USA, dpd@cs.princeton.edu
[2] Max-Planck-Institute für Informatik, Im Stadtwald,
66123 Saarbrücken, Germany, gunopulo@mpi-sb.mpg.de

Abstract. We present some problems with geometric characterizations that arise naturally in practical applications of machine learning. Our motivation comes from a well known machine learning problem, the problem of computing decision trees. Typically one is given a dataset of positive and negative points, and has to compute a decision tree that fits it. The points are in a low dimensional space, and the data are collected experimentally. Most practical solutions use heuristic algorithms.
To compute decision trees quickly, one has to solve optimization problems in one or more dimensions efficiently. In this paper we give geometric characterizations for these problems. We present a selection of algorithms for some of them. These algorithms are motivated from practice, and have been in many cases implemented and used as well. In addition, they are theoretically interesting, and typically employ sophisticated geometric techniques. Finally we present future research directions.

1 Introduction

To set up the discussion we give a description of decision trees and their uses. The problem of computing decision trees from example sets, an important problem on its own right, is a special case of a central problem in machine learning, the problem of learning concepts from examples. We review some strategies to deal with this problem.

Finally we present the minimizing disagreement problem. Fast algorithms for this problem are important for practical and theoretical reasons. For example, most applied algorithms for decision tree computation have to solve some version of this problem.

1.1 Decision Trees

Decision trees are a very useful tool in artificial intelligence where they are used to classify new examples and to summarize concepts (see [Q93], [BFOS], [SCFMW], [MKS], [Mo], [PP], [SL]). Decision trees have been used successfully in many practical applications, such as finding protein sequences or identifying cosmic objects in space images ([SCFMW]). They are also beginning to penetrate planning applications where they are used to compress mapping from states to actions, probabilities, and utilities ([BDG]).

A *decision tree* uses recursive partitioning to divide the instance space into mutually disjoint regions. Each partitioning is represented by a node in the tree, and is typically a split on one of the dimensions of the instance space. Each final region is assigned a single value and is represented by a leaf in the tree (Fig. 1).

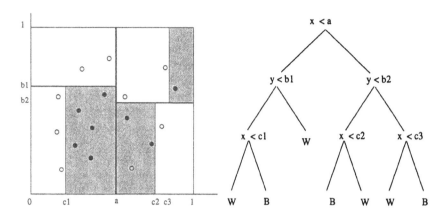

Fig. 1. A decision tree with depth 3. Black points are positive examples, and white points negative examples. The input examples have 2 attributes (x and y axis). Only one point is incorrectly classified.

When constructing decision trees, we typically wish to produce accurate and concise (small) trees. The standard algorithms build trees top down, recursively splitting the examples using tests on the attributes ([Q93],[BFOS], [MKSB].) The splitting stops when either all or most of the examples remaining belong to the same class.

For examples with real-valued attributes, the decision tree algorithm considers splitting the examples into two groups, and picks the best "cut point" available. Once a set of n examples is sorted along a particular attribute, an algorithm must consider only $n - 1$ places as possible cut points, corresponding to the midpoints between every successive pair of examples. This naturally leads to an efficient top-down algorithm for constructing binary decision trees. The amount of work required at each node of the tree is $O(dn)$ where d is the number of attributes (dimensions). An additional $O(dn \log n)$ is required to sort the examples along each dimension, but with careful bookkeeping this can be done just once for the whole tree. Thus the cost of building binary decision trees is just $O(dnT + dn \log n)$, where T is the tree size. Since in general $T = O(n)$, the cost of building a tree is $O(dn^2)$. A test (or "split") in such a tree has the form $X_i \geq a$ for attribute i.

Once the decision tree is computed, the time to classify an unknown input point is proportional to the height of the tree. For a given point in the instance space we follow a path from the root to a leaf. At each node we use the point's relevant coordinate to decide which subtree to follow. Since each leaf represents a uniform region, the point's classification as a positive or a negative example depends on the kind of region that we reach.

The important problem is to compute decision trees whose accuracy can be guaranteed. Such algorithms are especially needed for critical applications in which health and safety are involved; for example, medical diagnosis or air traffic control systems. There are two main challenges to producing algorithms that deliver high accuracy in practice: obtaining sufficient statistics from data and limiting the running time of the learning algorithm. That is, learning algorithms that provide high accuracy require large numbers of examples, which are often very hard to obtain. Even when sufficient examples are available, the algorithms themselves may be too slow to be used in any practical application.

Thus one promising short-term goal of machine learning research is to produce algorithms that can run fast (e.g., in time that is linear in the number of examples) and that also produce trees with guaranteed accuracy bounds. Ideally, we would like to produce a classifier that with high confidence has a low error probability. However, for computational efficiency considerations we often work with restricted methods such as linear classifiers. In these cases we would like to know that the classifier is as good as possible given the data, a notion quantified below.

1.2 Learning from Examples Using the Agnostic PAC Model

Computing decision trees is an instance of the following core problem in machine learning, the problem of learning a concept from examples. In this problem, we are given a set of examples, each labeled either as a positive example of the concept or as a negative example. These examples are typically points that lie in a low-dimension (around 10 in many cases) space. Each dimension represents an attribute of the concept, and can have values in a continuous or a discrete range. The problem is to find a hypothesis that represents the known examples, and hopefully the concept itself, as well as possible. Such a hypothesis can be used either as a predictor for new query points or as a classifier to the original training set, to intuitively describe the concept.

A major goal of machine learning is to produce practical and efficient learning algorithms that will provide reasonable guarantees on accuracy after seeing a very small number of points. The main difficulty is coming up with practically reasonable statistical assumptions (e.g, Bayesian priors) that can be used to derive strong confidence in a small probability of error after just a few samples. A promising intermediate goal is to produce practical and efficient algorithms that at least provide strong asymptotic guarantees.

The *agnostic PAC-Learning model*, introduced Haussler in [Hau], provides an important link between theoretical and applied machine learning. This model, a variant of the PAC-learning model (see [V]), provides an adequate format for the generic formulation of real-life classification problems because it does not assume any knowledge of the target concept on the part of the learner. In agnostic PAC-learning the training examples are generated by an arbitrary unknown distribution, and the learning algorithm is given a set of hypotheses $\mathcal{H}$. The learning algorithm is required to find a hypothesis in $\mathcal{H}$ that approximates the target almost as well as any other hypothesis in $\mathcal{H}$. An algorithm is an *efficient agnostic-PAC learner* for $\mathcal{H}$ if this is done in polynomial time.

Haussler ([Hau]) shows that if the VC dimension of the hypothesis class $\mathcal{H}$ is finite then there exists a polynomially bounded function m such that, with probability at least $1-\delta$, a hypothesis $H \in \mathcal{H}$ that minimizes the number of the incorrectly characterized examples for a training set S of $n = m(\text{VC-dim}(\mathcal{H}), 1/\delta, 1/\epsilon)$ examples, approximates the target distribution within an $1/\epsilon$ term of the best possible hypothesis in $\mathcal{H}$. So, if the VC dimension of $\mathcal{H}$ is finite, a polynomial time algorithm that finds the hypothesis in $\mathcal{H}$ that minimizes the error for a training set S gives an *efficient agnostic-PAC learner* for $\mathcal{H}$.

It follows that if the training set S is sufficiently large, a hypothesis in $\mathcal{H}$ that minimizes the error for S, is guaranteed to approximate the target distribution almost as well as the optimal hypothesis in $\mathcal{H}$.

1.3   The Minimizing Disagreement Problem

The results of §1.2 show that in order to prove a positive result for efficient agnostic PAC-learning with a specific hypothesis class $\mathcal{H}$ of bounded VC-dimension, it suffices to design a polynomial algorithm for a finite optimization problem, the *minimizing disagreement problem* for the hypothesis class $\mathcal{H}$. This is the problem of computing, for any given finite training sequence T of labeled points, some hypothesis $H \in \mathcal{H}$ whose *error* (the number of incorrectly characterized examples) is minimum among all hypotheses in $\mathcal{H}$. An algorithm that solves the minimizing disagreement problem for $\mathcal{H}$ is, together with the bounds for the minimum number of training examples given above, an agnostic PAC-learner for hypothesis class $\mathcal{H}$. By the same reasoning, an efficient ϵ_1-approximation algorithm for the minimizing disagreement problem for $\mathcal{H}$ can produce a hypothesis H with true error up to $\epsilon_1 + \epsilon$ from the optimal.

As we saw in §1.1, efficient algorithms for the minimizing disagreement problem are very desirable from the point of view of applied decision tree computation. The heuristic greedy algorithms that are used in practice usually choose at each step a split that minimizes the error. In most cases the splits are hyperplanes in one dimension because here the optimum split can be computed quickly (see §2.1). Experimental results however, show that more sophisticated splitting rules can offer substantial advantages, reducing the size of the decision tree, and increasing its accuracy. The experimental results are due to Weiss et al ([WK91]) and Lubinsky ([L]), who consider splits along simple geometric shapes (in particular axis oriented rectangles in two dimensions), and Murthy et al ([MKSB]), who use splits along a hyperplane in 2 or more dimensions of the input space. However sophisticated splitting rules can be used in practice only if there exist efficient algorithms that solve the corresponding minimizing disagreement problem.

In [DGM] the minimizing disagreement problem is shown to be equivalent to the problem of computing the maximum bichromatic discrepancy of a point set. We believe that this intuitive geometric characterization of the problem simplifies the design of efficient algorithms.

2 Known Results

A currently very active area of research in machine learning is to improve the accuracy of decision trees, and using better splitting rules seems to be a straight-forward way to achieve this. Another active area is to evaluate the practical capabilities of agnostic-PAC learners.

In this section we present the known results for the minimizing disagreement problem for geometric hypotheses. There has been a lot of theoretical work in computational learning theory on the PAC and agnostic-PAC learnability of various hypothesis classes ([KSS], [F], [K], [Ma], [DG]). Unfortunately much of this work is not practical either because of the complexity of the algorithms or because of the inadequate performance of the hypothesis classes. One important aspect of geometric concepts is that they are powerful enough to be useful, yet they are intuitive and simple enough that efficient algorithms can be designed.

In this section we concentrate on results that have been used in applications or are motivated by practical applications. The results that are described in the first two sections can be used in greedy algorithms for decision tree computation. The next two sections present the only known decision tree algorithms that are provably optimal and are also interesting from a practical standpoint. In most cases experimental data are also available. Most experiments use a set of standard datasets (available from ftp site ics.uci.edu, pub/machine-learning-databases) that have been obtained from real problems in the last few years. Each dataset contains a few hundred to a few thousand points, and the input domain has usually less than 10 dimensions (attributes).

2.1 Computing the Optimal Hyperplane

Computing the hyperplane that minimizes the error is an important subroutine in many greedy algorithms for decision tree computation. In most applications each split is along a single dimension. However Murthy et al ([MKSB]) present an algorithm where each split is a hyperplane in d dimensions (d is the dimension of the example domain). The chosen split is a local optimum, but their experiments demonstrate that the decision trees thus produced perform well.

We assume that we are given a training set S of n points in the d dimensional unit box ($S \subset [0,1]^d$.) Each point in the training set is positive or negative.

A hyperplane H divides the input space into two halfspaces, one of them labeled positive, and the other negative. Each point in S falls in one of the two halfspaces. The error of H is the number of points in S that are incorrectly classified. The problem is to find the hyperplane that minimizes the error.

In one dimension the example set consists of points on a single line, and the problem can be easily solved in $O(n \log n)$ time. Indeed we can find the split that minimizes the error with a single pass over the sorted points. On the other hand, Kearns et al ([KSS]) show that when the dimension of the example space is arbitrary, the problem is NP-complete.

[DGKFS] give an $O(n^2)$ static algorithm (see also [DEM]), and a $O(n^2 \log n)$ dynamic algorithm for points on the plane. This algorithm uses the dual arrangement of the point set, and finds the error for each hyperplane that passes through a pair of points. The smallest error is then reported (Fig. 2).

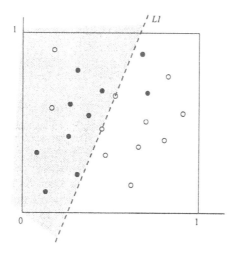

Fig. 2. A halfspace that minimizes the number of incorrectly characterized examples for the 2-D point set shown

2.2 Computing the Optimal Rectangle

Here we look at the minimizing disagreement problem when the class of hypotheses is the set $\mathcal{R}$ of axis aligned, but otherwise arbitrary, rectangles (Fig. 3).

Simple hypotheses classes of this type have turned out to be quite interesting in applied machine learning. Weiss et al ([WK90], [WGT], [WK91]) have shown through experiments that for many of the standard benchmark datasets a short rule that depends on only two of the attributes, and which is a boolean combination of expressions of the form "$a_j > c$" or "$a_j = c$", provides the best available prediction-rule. For example it is reported ([WK91]) that for the appendicitis dataset the complement of a rectangle (in 2 of the 8 attributes of the particular dataset) is the prediction rule that performs best. In addition, optimal hypotheses of a simple type have the additional advantage that they provide a human user valuable heuristic insight into the structure of a real-world learning problem. Lubinsky also studied the the problem of computing the optimal rectangle,

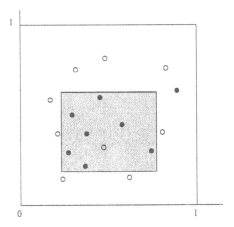

Fig. 3. The rectangle shown minimizes the error for all axis aligned 2-D rectangles

and gave a cubic algorithm for it ([L]). He gives experimental results that show that the use of this decision rule produces smaller decision trees on the average.

[DGM] give an $O(n^{2d-2} \log n)$ algorithm to compute the optimum rectangle in d dimensions. The main idea of the algorithm is to combine an efficient dynamic one dimensional algorithm and a plane sweep technique. The algorithm's simplicity allows easy implementation. Experiments also show that it is reasonably fast.

Fischer and Kwek ([F], [K]) consider the problem of computing the optimum k-gon in two dimensions (for a fixed k). They present similar dynamic algorithms to solve it. The fastest dynamic algorithm has a $O(kn^6)$ running time ([K]). For the same problem ([DG]) give a different algorithm with a running time of $O(n^{2k-1} \log n)$.

2.3 Decision Trees that Minimize the Error

In this section we to consider the hypothesis classes $\mathcal{T}(1, K)$ and $\mathcal{T}(2, K)$. A decision tree T is in $\mathcal{T}(2, K)$ if it has a binary split in one dimension, and a K-way split in a second dimension. T therefore has depth 2, and $2K$ leaves (K is a constant). Each leaf corresponds to a rectangular region which is labeled positive or negative. The instance space may have more than 2 dimensions, but at most 2 are used in the decision tree (Fig. 4).

A decision tree in $\mathcal{T}(1, K)$ is simply a K-way split in one dimension (Fig. 5). Holte ([Ho]) has shown that this simple classification method can be surprisingly effective. His experiments demonstrate the interesting fact that in many practical cases, even in instance spaces of more than 10 dimensions, there exists one dimension that fairly accurately describes the concept. [FKS] give an algorithm with $O(dKn + dn \log n)$ running time that computes the tree in $\mathcal{T}(1, K)$ that minimizes the error for a set of n points in d dimensions. They also give new experimental results that verify that these simple trees have good accuracy.

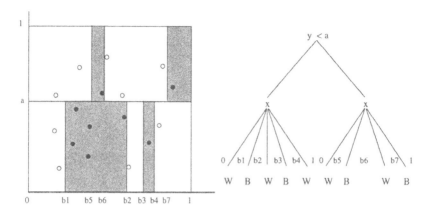

Fig. 4. An optimal decision tree in $\mathcal{T}(2,5)$.

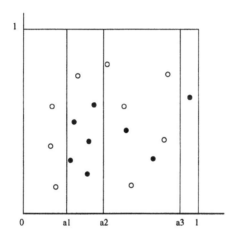

Fig. 5. An optimal decision tree in $\mathcal{T}(1,4)$. The input set has 2 dimensions.

Auer et al ([AHM]) were the first to give a practical algorithm for $\mathcal{T}(2,K)$. Given a set $S \subset [0,1]^d$ ($|S| = n$) of positive and negative examples, the tree in $\mathcal{T}(2,K)$ that minimizes the error can be computed in $O(d^2 K^2 n \log n)$ time.

This algorithm also combines a dynamic one-dimensional algorithm and a plane sweep. They also show that the VC-dimension of $\mathcal{T}(2,K)$ is finite if d is finite, and therefore this algorithm gives an efficient agnostic-PAC learner of $\mathcal{T}(2,K)$. The algorithm is simple to implement, and efficient. In their experiments they compare the accuracy performance of the $\mathcal{T}(2,K)$ class of decision trees with that of sophisticated heuristic algorithms and show that this class offers competitive performance with Q4.5 ([Q93]) on standard datasets.

Decision trees in the $\mathcal{T}(2,K)$ class can classify incorrectly many points of the training set, if the parameter K is set at too low a value. Ideally we would like to compute trees with small error. But to achieve that, K cannot be too small. To address this problem [DGK] present an algorithm that attempts to compute a decision tree with small error. Given a set $S \subset [0,1]^d$ ($|S| = n$) the tree in $\mathcal{T}(2, K)$ that minimizes the error can be computed in $O(d^2 \epsilon K n + d^2 K^2 n + d^2 n \log n)$ time if there exists a tree that incorrectly classifies at most ϵ points.

In a different approach to generalize the results of [Ho] and [FKS], [DG] introduce the hypothesis set $\mathcal{S}_K$ of K-stripes. A K-stripe is defined by K parallel lines of arbitrary orientation (Fig. 6). The following result is shown: The stripe in $\mathcal{S}_K$ that minimizes the error for a set $S \subset [0,1]^2$ ($|S| = n$) can be computed in $O(K^2 n^2 \log n)$ time, and in $O(n^2)$ space. This algorithm uses a plane sweep in the dual space arrangement.

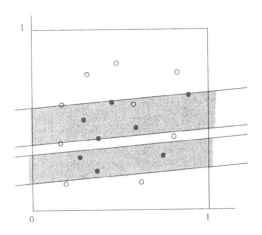

Fig. 6. An optimal 4-stripe.

2.4 Perfect Decision Trees

Let $\mathcal{T}_k$ be the set of decision trees with hyperplane splits, and depth at most k. For a given set of examples S, and a given constant k, a decision tree in $\mathcal{T}_k$ that classifies all points in S with no error is called a *perfect decision tree* for S and χ.

Decision trees of small depth are intuitive tools to describe and argue about concepts. Unfortunately, the greedy algorithms that are used in practice don't usually produce trees that are optimal in depth. However it is in many cases desirable to have algorithm that computes perfect decision trees of small depth. For example, an interesting idea is to first use a greedy algorithm, to get a decision tree T with small error. Then, to eliminate the examples that are not correctly characterized by T. T is then a perfect decision tree for the set of the remaining examples. Using an optimization algorithm on the new set we can obtain a new tree T' which has at most equal error to T, but is probably smaller and better balanced.

Das and Goodrich ([DaG]) prove that, given a set S of positive and negative points in three dimensions ($S \subset \Re^3$), it is NP-complete to compute a decision tree with at most k nodes that correctly classifies all $|S|$ points.

If however if we don't allow arbitrary hyperplane splits, and use splits parallel to the axes instead (that is, only allow one dimensional splits,) there exists a simple dynamic algorithm of $O(n^5)$ running time that finds the tree of minimum depth (see for example [DGKFS]). In addition, [DGKFS] show the following result: Given a set of points $S \in [0, 1]^d$ ($|S| = n$), and a constant k, a binary decision tree T that correctly classifies all points in S, can be found in in $O(d^k n \log^{k-2} n)$ time, if such a tree exists. In the special case of $d = 2$, the running time is reduced to $O(2^{2^3} n \log^{k-3} n)$. Preliminary experimental data show that this algorithm works well in some standard datasets.

Arkin et al ([AMMRS]) consider the following problem from computer vision: given a library S of k geometric models in two dimensions (typically simple polygons), and an image of a model, find the corresponding model in the library. They show that the problem of computing perfect decision trees of optimum depth, for geometric two dimensional objects, is NP-complete. However, they give a greedy algorithm that computes a perfect decision tree of depth $H = O(\log kh)$ in $O(Hn^2)$ time (where h is the depth of the optimum tree).

3   Conclusion

The algorithms we survey are directly motivated by machine learning problems. In some cases these algorithms have been applied to practical problems ([SCFMW], [Q93], [WK91]). Since many concepts in standard benchmarks can be effectively learned by considering just a few relevant dimensions, the optimal algorithms presented here can be used in practice ([Ho], [AHM], [DGK], [L]). Many problems remain open in this area, and we give some of the most interesting ones:

An important research goal is to investigate which hypothesis classes are expressive enough to be useful, yet simple enough so that fast and practical algorithms to compute accurate classifiers can be designed. In practical terms this means that such algorithms should have running times of $O(n\operatorname{poly}\log n)$, and use relatively simple data structures. Examples of such hypothesis classes are low depth decision trees, and x-monotone regions ([ACKT]).

Most of the known algorithms are not provably optimal, so there is scope for improving the running times. Also, approximation algorithms should be investigated. In many cases faster running times are more important than accuracy.

A different approach that has not been theoretically investigated so far is to use randomization techniques to compute approximate solutions. In practice (see for example [MKSB]) randomized algorithms give satisfactory solutions, but no performance guarantees are given. There is experimental evidence to suggest that randomized algorithms could be particularly effective when the user has some apriori knowledge about the target concept.

An important problem not considered so far is the statistical motivation for computing the smallest decision tree for a given training set. It has been shown in [Q95] that "oversearching" can in fact hurt the accuracy of practical learning algorithm. We are actively investigating this subtle problem in the context of the bias/variance trade-off and other issues.

Another goal is to implementing the algorithms and performing experiments. More detailed experimental study would help determine how useful the optimal algorithms can be in practice, and how their performance compares with the heuristic algorithms currently in widespread use, and in addition provide insights on the algorithm development in the field.

Many implementation issues have also to be addressed. For example, most databases contain errors (for example having the same point appear twice, once as a positive and once as a negative example), missing values, and many special cases (a typical problem is having too many points with an identical coordinate, or lying on a single line). The implemented algorithms should be able to handle all these cases correctly, and without performance degradation.

References

[AMMRS] E. Arkin, H. Meijer, J. Mitchell, D. Rappaport, and S. Skiena, Decision Trees for Geometric Models. *Proc. Comput. Geom. Conf.* (1993), 369-378.

[ACKT] T. Asano, D. Chen, N. Katoh, and T. Tokuyama, Polynomial-Time solutions to Image Segmentation. *Proc. 7th ACM-SIAM Symp. on Disc. Algorithms* (1996), 104-113.

[AHM] P. Auer, R. Holte and W. Maass, Theory and Applications of Agnostic PAC-Learning with Small Decision Trees. *Proc. 12th Int. Conf. Machine Learning* (1995).

[BDG] C. Boutilier, R. Dearden and M. Goldszmidt. Exploiting Structure in Policy Construction *Proceedings of the International Joint Conference on Artificial Intelligence*, Montreal, Michigan, 1995.

[BFOS] L. Breiman, J.H. Friedman, R.A. Olshen, and C.J. Stone. *Classification and Regression Trees*, Belmont, CA: Wadsworth International Group, 1984.

[BN] W. Buntine and T. Niblett, A further comparison of splitting rules for decision-tree induction. *Machine Learning*, 8 (1992), 75-82.

[DaG] G. Das and M. Goodrich, On the complexity of Optimization Problems for 3-Dimensional Convex Polyhedra and Decision Trees. *WADS 1995*.

[DEM] D. Dobkin, D. Eppstein and D. Mitchell, Computing the Discrepancy with Applications to Supersampling Patterns. *ACM Transactions on Graphics*, to appear.

[DG] D. Dobkin and D. Gunopulos, Concept Learning with Geometric Hypotheses, *8th ACM Conference on Learning Theory* (1995).

[DGK] D. Dobkin, D. Gunopulos and S. Kasif, Computing optimum shallow decision trees. *4th AI and Math. Symposium* (1996).

[DGKFS] D. Dobkin, D. Gunopulos, S. Kasif, J. Fulton and S. Salzberg, Induction of Shallow Decision Trees. submitted to *IEEE PAMI*.

[DGM] D. Dobkin, D. Gunopulos and W. Maass, Computing the maximum Bichromatic Discrepancy, with applications in Computer Graphics and Machine Learning. *J. Comp. Syst. Sciences*, to appear.

[F]   P. Fischer, More or less efficient agnostic learning of convex polygons. *8th ACM Conference on Computational Learning Theory* (1995).

[FKS]   T. Fulton, S. Kasif and S. Salzberg, An Efficient Algorithm for Finding Multi-way Splits in Decision Trees. *Proc. Machine Learning 1995*.

[Hau]   D. Haussler, Decision theoretic generations of the PAC-model for neural nets and other applications. *Inf. and Comp.*, 100 (1992), 78-150.

[Ho]   R.C. Holte, Very simple classification rules perform well on most commonly used datasets. *Machine Learning*, 11 (1993), 63-91.

[KSS]   M. Kearns, R.E. Schapire and L.M. Sellie, Toward efficient agnostic learning. *5th ACM Workshop on Computational Learning Theory* (1992), 341-352.

[K]   S. Kwek, Minimizing disagreements for geometric regions using dynamic programming, with applications to machine learning and computer graphics. Manuscript, 1995.

[L]   D. Lubinsky, *Bivariate splits and consistent split criteria in dichotomous classification trees*. Ph.D. Thesis, Rutgers University, Department of Computer Science, 1994.

[Ma]   W. Maass, Efficient Agnostic PAC-Learning with Simple Hypotheses. *7th Ann. ACM Conference on Computational Learning Theory* (1994), 67-75.

[Min]   J. Mingers, An empirical comparison of pruning methods for decision tree induction. *Machine Learning, 4* (1989), 227-243.

[Mo]   B. M.E. Moret, Decision Trees and diagrams. *Computing surveys, 14(4)* (1982), 593-623.

[MKS]   S. Murthy, S. Kasif and S. Salzberg, A system for induction of oblique decision trees. *Journal of Artificial Intelligence Research, 1* (1994), 257-275.

[MKSB]   S. Murthy, S. Kasif, S. Salzberg and R. Beigel, OC1: Randomized induction of oblique decision trees. *AAAI 93* [2], 322-327.

[PP]   R. W. Payne and D. A. Preece, Identification trees and diagnostic tables: A review. *Journal of the Royal Statistical Society: series A, 143* (1980), 253.

[Q86]   J.R. Quinlan. Induction of Decision Trees. *Machine Learning, 1* (1986), 81-106.

[Q93]   J.R. Quinlan. *C4.5: Programs for Machine Learning*, Morgan Kaufmann, Los Altos, CA, 1993.

[Q95]   J. R. Quinlan. Oversearching and Layered Search in Empirical Learning. *Proceedings of the International Joint Conference on Artificial Intelligence*, Montreal, Michigan, 1995.

[SL]   S. R. Safarin and D. Landgrebe, A survey of decision tree classifier methodology. *IEEE Transactions on Systems, Man and Cybernetics, 21(3)* (1994), 309-318.

[SCFMW]   S. Salzberg, R. Chandar, H. Ford, S. Murthy and R. White, Decision trees for automated identification of cosmic-ray hits in *humble space telescope* images. *Publications of the Astronomical Society of the Pacific, 107*, 1-10 (March 1995).

[V]   L.G. Valiant, A theory of the learnable. *Comm. of the ACM* 27 (1984), 1134-1142.

[WGT]   S.M. Weiss, R. Galen and P.V. Tadepalli, Maximizing the predictive value of production rules. *Art. Int.* 45 (1990), 47-71.

[WK90]   S.M. Weiss and I. Kapouleas, An empirical comparison of pattern recognition, neural nets, and machine learning classification methods. *11th Int. Joint Conf. on Art. Int.* (1990), Morgan Kauffmann, 781-787.

[WK91]   S.M. Weiss and C.A. Kulikowski, *Computer Systems that Learn*, Morgan Kauffmann Publishers, Palo Alto, CA, 1991.

Matching Convex Polygons and Polyhedra, Allowing for Occlusion

Ronen Basri[*1] and David Jacobs[2]

[1] Dept. of Applied Math., The Weizmann Inst. of Science, Rehovot, 76100, Israel
[2] NEC Research Institute, 4 Independence Way, Princeton, NJ 08540, USA

Abstract. We review our recent results on visual object recognition and reconstruction allowing for occlusion. Our scheme uses matches between convex parts of objects in the model and image to determine structure and pose, without relying on specific correspondences between local or global geometric features of the objects. We provide results determining the minimal number of regions required to uniquely determine the pose under a variety of situations, and also showing that, depending on the situation, the problem of determining pose may be a convex optimization problem that is efficiently solved, or it may be a non-convex optimization problem which has no known, efficient solution. We also relate the problem of determining pose using region matching to the problem of finding the transformation that places one polygon inside another and the problem of finding a line that intersects each of a set of 3-D volumes.

1 Introduction

In this paper we will review a series of our recent results, with collaborators, on visual object recognition and reconstruction that seem particularly relevant to the computational geometry community. These results are given in more detail in [8, 9, 22, 25]. This work uses matches between parts of objects to determine structure and pose, without relying on specific correspondences between local or global geometric features of the objects.

As an example of the type of information that we use, suppose one has matched the seat of a chair in a 3-D CAD model to a region in an image that looks like the chair seat, perhaps because it has a similar color or texture. In the image, the chair may be partly occluded by intervening objects, or by itself. If the extent of such occlusion is unknown, such a match still provides a strong constraint on the possible poses of the model in the scene; to be consistent with this match, a pose must result in the model of the chair seat projecting into the image so that it completely accounts for the corresponding region in the image.

Our work explores how to make use of this type of region matching to determine the pose of an object, in order to recognize it. There are two main

---

* The research of Ronen Basri was supported in part by the Israeli Ministry of Science, Grant No. 6281 and in part by the Unites States-Israel Binational Science Foundation, Grant No. 94-00100.

questions that arise: when do matches between regions provide enough information to adequately determine pose? And when can we efficiently find poses that are consistent with a set of region matches? The answers to these questions vary depending on the type of model we consider (3-D or 2-D), the type of viewing transformation we allow for (similarity, affine or perspective), the number of regions that have been matched, and the amount and type of occlusion for which we allow. We have analytically determined answers to a number of these questions, showing the number of regions that must be matched to uniquely determine pose under a variety of situations, and also showing that depending on the situation, the problem of determining pose may be a convex optimization problem that is efficiently solved, or it may be a non-convex optimization problem which has no known, efficient solution.

We will also describe how the problem of determining pose using region matching is closely related to two problems that have been studied in the computational geometry community. The first is the problem of finding the transformation that places one polygon inside another. This has been considered in several papers ([13, 14, 32]), and we will in particular discuss the relationship between our work and recent work that considers this problem when the inner polygon may be scaled ([3]). The second problem is that of finding a line that intersects each of a set of 3-D volumes (i.e., a *line traversal*). Results on this problem have been given by [4, 5, 17, 24]. We show that this problem is closely related to that of determining the pose of a 3-D object using a 2-D image, and build on some of these results in our work. We hope, therefore, to spur additional interest in these problems within the computational geometry community by pointing out their close connection to a significant application in computer vision.

2   Previous Approaches to Object Recognition

One very common approach to visual object recognition is to hypothesize correspondences between portions of a known model and portions of an image, use these correspondences to determine the pose of the model in the scene, and then look for additional evidence in the image to confirm or reject this hypothesized model pose (e.g., [1, 2, 16, 18, 21, 33]). Within this paradigm, the problem of pose determination is central, and this is the problem on which we focus in this paper. Moreover, pose determination may be an important problem in its own right, when a human operator manually specifies correspondences between a model and image that are assumed correct, and the relative position of the camera and scene must be determined.

One of the novel aspects of our work is the type of information that we use to determine pose. Current methods for pose estimation rely primarily on three different types of descriptions of model and image information to determine pose. One type of method uses *global* properties of the objects. These properties may include for example the position of the center of mass of the object and its principal axes. The corresponding properties of the object in the image are then computed and used to bring the object into the canonical description (e.g., [15, 19, 27, 29, 31, 28]). This approach is computationally efficient, since processing the image can be carried out independently of the model. However, it requires a good segmentation of the object, and it is sensitive to occlusion, and in particular to self occlusion. This makes the method unsuitable for recognizing 3-D objects from single 2-D images, or to recognition in cluttered scenes.

Another type of method uses *local* properties of the objects, such as corners, points of curvature extrema, and straight line segments (e.g., [1, 2, 16, 18, 21, 33]). Matching subsets of the model and image features are sufficient to recover a transformation that relates the two. Since only a subset of the features need to be matched, local methods can recover the pose of objects in the presence of clutter and occlusion. These methods, however, suffer from several limitations. First, local features may vary considerably due to changes in illumination and viewpoint. In particular, local features are often unsuitable for recognizing smooth curved objects. Secondly, local methods can be computationally expensive. For instance, for polyhedral objects all triplets of model points must be tried against all triplets of image points to guarantee that a solution is found ($m^3 n^3$ matches).

Both of these two methods have been difficult to apply to the problem of recognizing smooth 3-D objects. Global methods are not suitable in this situation due to self-occlusion and viewpoint dependent effects; for example the center of mass of the 2-D image of an object need not be the projection of its 3-D center of mass. On the other hand, features derived from the boundary of an object are typically highly sensitive to viewpoint, because viewpoint determines which portions of an object's surface actually generate its boundary in the image. Therefore, recognition of such objects is typically attempted by computing more complex properties of the parts of 3-D objects. A typical example of this is the large body of recognition work using generalized cylinders (eg. [11, 12, 23]). In this case, a 2-D image of an object part is used to determine a complex algebraic description of the 3-D model part, including its axis and a rule that describes how its cross-section varies along this axis. Other approaches include *geons* and *superquadrics* [7, 10, 26]. These parts-based recognition methods place a large burden on bottom-up processing, and it has proven difficult to compute rich and accurate descriptions that characterize the 3-D structure of parts, especially when the corresponding image regions may be noisily detected with partial occlusions. However, these approaches offer the promise of recognizing objects by indexing into a large data base of possible objects, because the model description derived from the image is so rich in information.

Our work also relies on a part-based description of an object. We use a much simpler description of the 2-D image and 3-D model parts, however, and focus on pose determination rather than indexing. Our method describes image and model parts simply as sets of points, without demanding the construction of a more complex representation. One of the main contributions of our work, then is to show that we can perform part-based recognition using a very simple representation.

3 Problem Statement: Pose Determination

First we consider the case in which we must determine the pose of a known model using possibly occluded regions in a 2-D image. We assume that a hypothesized match exists between a set of model volumes and image regions. We further assume that the model consists of a set of 2-D or 3-D convex volumes denoted by: $V_1, ..., V_k$. Similarly, we assume that the image consists of 2-D regions, which are each subsets of $\mathcal{R}^2$, and which we denote by: $R_1, ..., R_k \subset \mathcal{R}^2$. Our solution methods will apply to the case where the model and image sets are convex; if we wish to make use of non-convex volumes or regions we should first take their convex hulls. Our methods can naturally apply also when some or all of the correspondences are between point features, or (possibly partially occluded) line segments, since these are convex.

Next, we suppose that the image was generated by applying some transformation, T, that maps points in the model to points in the image. For 2-D models we consider similarity, affine or projective transformations. For 3-D models, we allow for a 3-D affine transformation to position the model in the scene, and then orthographic or perspective projection to create an image. We denote a point in model space by $\mathbf{p} = (x, y, z)$ if the point is 3-D, or by $\mathbf{p} = (x, y)$ if it is 2-D. We denote a point in image space by $\mathbf{q} = (u, v)$. If $\mathbf{q} = T(\mathbf{p})$ then we denote $u = T_u(\mathbf{p})$ and $v = T_v(\mathbf{p})$.

We next consider three possible sets of constraints that might be applied, depending on the assumptions that are made with regard to occlusion. First, if we know that the image regions are completely unoccluded, we would ideally like to find a model transformation that matches each model volume perfectly to each image region (e.g., $R_i = TV_i$ for all $1 \leq i \leq k$). The set of solutions to this problem is clearly non-convex, however. For example, there are four distinct, disconnected transformations that match a square perfectly to a square for any of the classes of transformations we consider. This consequently appears to be a difficult problem to solve efficiently (although see [3] for insight into this problem when a similarity transformation is considered). For this reason and because it is important to allow for possible occlusions, we divide the set of constraints given to us by region matches into two classes, each of which in some cases lead to convex optimization problems.

The *backward constraints* require that every point in each image region be explained as the projection of some point in the corresponding model volume (i.e., $R_i \subseteq TV_i$, for all $1 \leq i \leq k$). This expresses our state of knowledge exactly when image regions are assumed to be arbitrarily occluded instances of the transformed model volumes.

It will also be of interest to consider the *forward constraints*. These require that every point in a model volume be explained by the corresponding image region (i.e., that $TV_i \subseteq R_i$, for all $1 \leq i \leq k$). These constraints exactly capture our state of knowledge in the somewhat contrived case where image regions are unoccluded, but due to undersegmentation of the image each image region is a superset of the projected model volumes. More significantly, the forward constraints are a true, though incomplete description of the situation in which image regions are not occluded, and we are able to show that in some cases we can find poses satisfying the forward constraints much more efficiently than those satisfying the backward constraints. Moreover, as we will discuss, we may apply the forward constraints in situations in which image regions may be occluded, but the boundaries of this occlusion have been identified.

When we wish to refer generically to either the forward or the backward constraints, we will call these *one-way constraints*. We now describe our results in applying the one-way constraints in the case of 2-D models.

4 Results for 2-D Models

If V_i and R_i are 2-D, and approximated by convex polygons, we may reduce the one-way constraints to a series of linear inequalities for the cases of similarity, affine or projective transformations. We show this in detail [8]; here we provide only a quick intuitive argument. The forward constraints require that the transformation map each model volume inside the corresponding image region. Because the 2-D transformations we consider are invertible, the backward constraints similarly require that the inverse of the unknown transformation map each image region inside the corresponding model volume. Since the volumes and regions are convex, the forward (or backward) constraints can be expressed by requiring that the unknown transformation (or its inverse) map each vertex of the model volume (or image region) to the appropriate side of each line bounding the corresponding image region (or model volume). E.g., the forward constraints have the form:

$$AT_u(\mathbf{p}) + BT_v(\mathbf{p}) + C \geq 0 \ .$$

As described in [8], these inequalities can be made linear for the transformations we consider. We therefore can find a pose which explains every image point with a corresponding model point using linear programming.

When only one model volume is matched to an image region, this method will not produce solutions of interest, however. The difficulty is that there are many transformations that satisfy the one-way constraints; when we allow for arbitrary occlusion or oversegmentation, and use only one-way constraints, we simply do not have enough information to find the transformation that actually produced the image regions. However, we show that we can determine a unique, correct transformation when we match more model volumes to image regions. We state the problem as follows:

Problem statement. We are given a set of regions produced by applying either a similarity, affine or projective transformation to a set of convex volumes. Under what circumstances is the similarity (affine, projective) transformations that will satisfy the one-way constraints uniquely determined? We will use notation appropriate for the forward constraints, but the problem is identical when either set of one-way constraints are considered.

We first show the following useful Lemma.

Lemma 1. *Let $V_1, V_2, ..., V_k \subseteq \mathcal{R}^2$ be k distinct (non-intersecting) volumes. Let $\mathcal{T}$ be the group of similarity, affine, or projective transformations. Let $R_i = T(V_i) \subseteq \mathcal{R}^2$, $1 \leq i \leq k$ be k regions obtained from $V_1, ..., V_k$ by applying an invertible transformation $T \in \mathcal{T}$. Then, there exists a transformation $T' \neq T, T' \in \mathcal{T}$ such that $T'(V_i) \subseteq R_i$, $1 \leq i \leq k$, if and only if there exists a transformation $\tilde{T} \neq I, \tilde{T} \in \mathcal{T}$ (I denotes the identity transformation) such that $\tilde{T}(V_i) \subseteq V_i$ for all $1 \leq i \leq k$.*

This shows that the solution to the uniqueness problem depends only on the structure of the model volumes, and is independent of the particular projection that produced the image. That is, without loss of generality we may assume that the image regions are identical to the model volumes. Using this, we show the following theorems:

Theorem 2. *Let $V_1, V_2 \subseteq \mathcal{R}^2$ be two distinct convex closed volumes ($V_1 \cap V_2 = \emptyset$). Then, the solution to the one-way matching problem with these volumes as a model under a similarity transformation is unique.*

Theorem 3. *Let $V_1, V_2 \subseteq \mathcal{R}^2$ be two distinct convex closed volumes. Then, the solution to the one-way matching problem with these volumes as a model under an affine transformation is not unique if and only if there exists a line l through V_1 and V_2 and a direction $\mathbf{v}$ such that contracting V_1, V_2 in the direction $\mathbf{v}$ toward l (denoted by $T_{l,\mathbf{v}}$) implies*

$$T_{l,\mathbf{v}}(V_i) \subset V_i \qquad i = 1, 2 .$$

Theorem 4. *Let $V_1, V_2, V_3 \subseteq \mathcal{R}^2$ be three convex closed volumes with non-zero areas such that there exists no straight line passing through all three volumes. Then, the solution to the one-way matching problem with these volumes as a model under a projective transformation is unique.*

(These are proven in [8, 9]).

We now illustrate these results with some examples using real images. These experiments are described in greater detail in [8, 9]. In brief, we automatically extracted convex groups of edges from images of an object, and then selected groups by hand to use as a model, and to match to this model in novel images. Figure 1 shows an image of a diskette used as a model, and Figures 2 and 3 show the results of matching this model to another image of the diskette by solving for a similarity and for an affine transformation using all five regions, or using only two regions. In both cases we have enough information theoretically to determine a unique pose, and in practice we produce an accurate match in spite of sensing errors.

These results show that in general, a match between two or three model and image polygons will be sufficient to determine the correct model pose. We note that these results show that a method that allows for arbitrary occlusion of image regions by using the backward constraints will produce correct results when there is in fact no occlusion. These results are significant for two reasons. First,

Fig. 1. An image of a computer diskette used as a model.

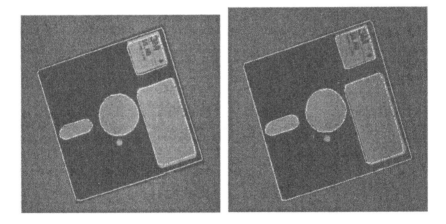

Fig. 2. Matching the diskette model to a novel image of the diskette under similarity (left) and affine (right) transformations. In each case the model regions' positions are indicated by white lines, overlaid on top of the image.

they show that our method will produce correct results when many volumes are matched, provided that two or three of them are unoccluded, even though the method does not require knowledge of which regions are unoccluded, and the extent of the occlusion that is present. This is because adding additional correspondences involving occluded image regions will only add correct (but possibly not useful) constraints to our linear program. Second, these results are precursors to results we will now discuss, which show that we may obtain unique solutions even when all image regions have significant occlusion.

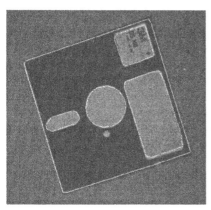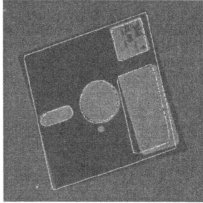

Fig. 3. Matching the diskette model to a novel image of the diskette using two regions only. Left figure: similarity. Right: affine.

5   2-D Matching with Occlusion

Our results above indicate that the backward constraints generally determine a unique affine or similarity transformation when two image regions are completely visible. However, not all the backward constraints produced by two matches are necessary to determine the solution uniquely. For example, when we seek an affine solution, which has six degrees of freedom, in principle, as few as seven linear constraints may suffice to determine a unique transformation. In practice, many of the constraints we derive are redundant.

This suggests that in the presence of occlusion the pose of the object may still be determined by the backward constraints even when only a small number of regions are used. This is because occlusion has the effect of replacing constraints that may tightly bound the solution with looser constraints. Still if enough constraints are unaffected by occlusion, a unique pose may still result.

The sensitivity of the backward constraints to occlusion, for affine and similarity transformations, is investigated in [25]. We review these results here. Because of Lemma 1 it is sufficient to consider only model and image pairs that are identical except for the occlusion. For such pairs we would like to determine whether the solution set to the backward constraints includes any transformation besides the identity. By analyzing polygonal regions it can be shown that whenever one region is entirely visible, the second region must contain a *critical point*, such that whenever this critical point is visible any transformation that is consistent with the backward constraints will have a pointwise-fixed line through the critical point. Such a transformation will contract the two regions in parallel directions toward the fixed line, where the direction of contraction is determined by the visible region.

An example for this situation is given in Figure 4. Because of the left triangle the two regions must contract in vertical directions. However, this contraction is not possible because points which lie along the portion of the side $\overline{p_5 p_6}$ above the fixed line will exit the triangle due to the contraction, violating the backward constraints. Thus, for this example, the solution to the backward constraints is unique as long as the critical point and at least one point on the upper portion of the side $\overline{p_5 p_6}$ are visible. As a result significant occlusions of the triangle which cover both vertices p_5 and p_6 will not affect the solution to the backward constraints. Similarly, it has been shown that even when the critical point is occluded the region can undergo significant occlusions and not affect the solution to the backward constraints. Simulations reveal that on average it is possible cover 60%-70% of the area of one region and still maintain the uniqueness of the solution (see Figure 5).

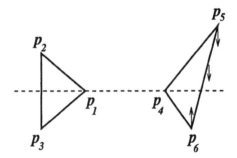

Fig. 4. Two triangles. p_4 is a critical point. The dashed line through p_1 and p_4 is the pointwise-fixed line. The arrows indicate the direction of contraction. Only affine transformations of this form will satisfy the backward constraints when the first triangle and p_4 are unoccluded.

6   Results for 3-D Models

In [22] we consider the problem of pose determination when our model volumes consist of 3-D convex polyhedra. Unlike the 2-D case, the forward and backward constraints must be handled quite differently, because the 3-D to 2-D viewing transformation is not invertible, and there is no longer a symmetry between the role of the model and the role of the image.

The forward constraints can be applied in much the same way for 3-D objects as for 2-D. We seek a 3-D affine transformation of the model volumes, such that every vertex of the model polyhedra projects inside the corresponding image polygon. Again, we may write this as a series of linear constraints, solvable with linear programming. And we are able to show analogous uniqueness results. In particular, in [22] we prove:

Theorem 5. *Suppose the transformation T maps the four model volumes, $V_0, V_1,$ V_2, V_3 to the four image regions, R_0, R_1, R_2, R_3, and there does not exist a plane*

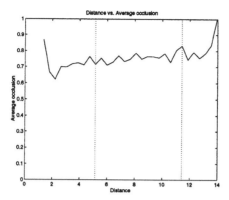

Fig. 5. Simulation results: the fraction of one triangle that may be occluded, while still permitting only the correct transformation, shown as a function of the distance between the two triangles. The range between the vertical lines contains the points for which significant amount of data was computed.

that intersects all four volumes. Then, T is the only transformation that satisfies the forward constraints.

We also show that except for special cases, three correspondences between model volumes and image regions suffice to determine a unique transformation, while two correspondences are not typically sufficient.

The forward constraints may be applied to allow for occlusion, when this occlusion can be identified in the image. If portions of the boundary of an image region are known to be due to occlusion, rather than truly indicating the bounding contour of the image region, these sides of the image polygon may be excluded from our linear program. This effectively constrains the model volume to project inside the largest possible convex polygon consistent with the information we have about the image region.

However, to allow for arbitrary, unidentified occlusions in the image we would really like to find transformations that satisfy the backward constraints, i.e., that project the model volumes so that they completely enclose the corresponding image regions. However, this turns out to be computationally difficult.

To better understand the complexity of this problem, we first point out that the *line traversal* problem is a subset of the problem of finding transformations that satisfy the backward constraints. To satisfy the backward constraints, we seek a transformation, T, such that $R_i \subseteq TV_i$ ($\forall i$, such that $1 \leq i \leq k$). Consider the special case in which every region consists of the same single point, q. It is possible to find a projection of the model volumes so that they all include the same point if and only if it is possible to find a line that intersects all the model volumes, since every image point is the projection of a line in 3-D. This is the problem of finding a line traversal to a set of volumes. Figure 6 illustrates this argument.

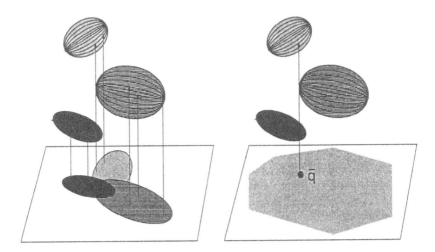

Fig. 6. On the left, we show three model volumes projecting to three, overlapping image regions. On the right, we imagine that these regions are occluded, except for one point which they all share. In this case, the problem of solving the backward constraints is equivalent to that of finding a line traversal of the model volumes.

[5] has shown that finding a line traversal to arbitrary 3-D volumes is not an LP type problem, because it can contain multiple solutions that are disconnected in the space of all possible lines. This shows that the problem of finding poses that satisfy the backward constraints is fundamentally more complex for 3-D models than it is for 2-D models.

However, [4] has shown that the line traversal problem is convex for an interesting special case, in which the 3-D volumes are axial rectanguloids. [24] has shown that this problem can in fact be solved using linear programming. We adapt this solution to our problem, and show that we may find poses that satisfy the backward constraints using linear programming, provided that we approximate the model volumes with axial rectanguloids. Figure 7 shows one example of our experiments with 3-D objects.

7   Conclusions

Much work on object recognition has assumed correspondence of local geometric features, however these methods have been difficult to apply to smooth curved objects. We have pursued a parts-based approach to visual object recognition that is also able to make use of geometric features such as points or lines when they are available. Typically, parts-based approaches assume little or no occlusion, so that either global properties of the parts may be used (e.g., moments of inertia) or so that rich, algebraic descriptions of the parts may be computed based on their boundary. Unlike these approaches, we compute pose using a very

Fig. 7. These figures show the performance of the system on realistic objects. On the left, the system uses the forward constraints to accurately determine the pose of the soda can. Four regions are used in this case. The regions are surface markings on the cylinder of the can, and a circular region from the top of the can. On the right, the approximate backward constraints are used to locate the pen box. Note that the soda can is occluding some of the regions used. The projected model volumes are shown in white outline, superimposed on top of the image.

simple description of the parts, and allow for occlusion in our problem formulation. We then consider basic questions about when a correspondence between possibly occluded object parts is adequate to determine an object's pose, and about the complexity of determining this pose.

For the case of planar objects we show that if as few as one object part is unoccluded, we can in many cases efficiently determine correct object pose, without knowing which parts are occluded, and to what extent. In the case of 3-D objects, we show that computing pose while allowing for arbitrary occlusion is fundamentally more difficult, and we have no efficient solution to this problem. We do, however, find efficient solutions when the model parts are approximated by axial boxes, or when the locus of occlusions of image parts have been identified in the image.

It is clear that two problems addressed in the computational geometry community are also central to understanding parts-based object recognition. One is the matching problem of finding transformations that map one object near, or inside of another. One weakness to our approach is that when the regions that are matched do not uniquely determine a solution, we do not provide a good method for finding the best of many possible solutions. Agarwal, Amenta, Aronov and Sharir[3] have recently provided one possible solution to this difficulty. Their algorithm finds the similarity transformation that places one convex polygon inside another, while scaling it to the maximum possible area. As this is not an LP-type problem, however, the complexity of the solution is somewhat greater than for our LP solutions. Interesting future work remains both in attempting to improve on the complexity of their algorithm, and to extend their results to other transformations. Open questions exist also as to whether one can improve upon our uniqueness results. We do not know under what circumstances we may be able to find a unique model pose matching two 3-D volumes to two image regions.

Another approach to matching object parts pursued by computational geometers involves minimizing their Hausdorff distance (e.g., [6, 20, 30]). This is a valuable approach to finding transformations that best fit two parts, although our work is significantly different in the way in which we allow for occlusion, and in our treatment of the projection of 3-D objects into a 2-D image. Open questions remain as to whether this can be done efficiently for a wide class of possible model transformations. [6] has shown that Hausdorff matching can be done more efficiently when object parts are assumed to be convex, for the transformation class consisting of translation plus scaling, but it has not been shown that this assumption can aid when more complex transformations are allowed.

The connection between our work and the line traversal problem also raises interesting questions. First, can other line traversal techniques for more general shapes than axial rectanguloids be applied to the problem of solving the backward constraints for 3-D models? It may be possible to adapt non-linear programming based algorithms for the line traversal problem to our matching problem. It may also be possible to find other limited line traversal problems that can be efficiently solved, and apply these to finding more accurate approximate solutions to our problem.

References

1. Alter, T. D., and W. E. L. Grimson, 1993, "Fast and Robust 3D Recognition by Alignment," in *Proc. Fourth Inter. Conf. Computer Vision*: 113–120.
2. Alter, T. D. and D. Jacobs, 1994, "Error Propagation in Full 3D-from-2D Object Recognition," *IEEE Conf. on Computer Vision and Pattern Recognition*: 892–898.
3. Pankaj Agarwal, Nina Amenta, Boris Aronov and Micha Sharir. "Largest placements and motion planning of a convex polygon," submitted to *Discrete and Computational Geometry*.
4. Amenta, N., 1992, "Finding a Line Traversal of Axial Objects in Three Dimensions," *Proc. of the Third ACM-SIAM Symp. on Discrete Alg.*:66-71.
5. Amenta, N., personal communication.
6. Amenta, N., 1994, "Bounded boxes, Hausdorff distance, and a new proof of an interesting Helly-type theorem," *Proceedings of the 10th Annual ACM Symposium on Computational Geometry*: 340–347.
7. Bajcsy, R. and Solina, F., 1987. "Three dimensional object representation revisited," *Proc. of The First Int. Conf. on Computer Vision, London*: 231–240.
8. R. Basri and D. Jacobs, 1995, "Recognition Using Region Correspondences," *Int. Conf. on Comp. Vis*:8–15.
9. R. Basri and D. Jacobs, 1995, "Recognition Using Region Correspondences," *The Weizmann Inst. of Science, TR-CS95-33*.
10. Biederman, I. 1985. "Human image understanding: recent research and a theory," *Computer Vision, Graphics, and Image Processing*, **32**: 29–73.

11. Binford, T.O., 1971. "Visual perception by computer," *IEEE Conf. on Systems and Control.*

12. Brooks, R., 1981. "Symbolic reasoning among 3-dimensional models and 2-dimensional images," *Artificial Intelligence,* **17**: 285–349.

13. B. Chazelle, 1983, "The Polygon Containment Problem," in *Advances in Computing Research, Vol. 1: Computational Geometry* (F. Preparata, Ed.):1–33, JAI Press, London England.

14. L. Chew and K. Kedem, 1993, "A Convex Polygon Among Polygonal Obstacles: Placement and High-Clearance Motion," *Comput. Geom. Theory Appls.* **3**(2):59–89.

15. Dudani S.A., Breeding K.J., and McGhee R.B., 1977. "Aircraft identification by moments invariants," *IEEE Transactions on Computations,* bf C-26(1): 39–46.

16. Fischler, M.A. and Bolles, R.C., 1981. "Random sample consensus: a paradigm for model fitting with application to image analysis and automated cartography," *Com. of the A.C.M.* **24**(6): 381–395.

17. Hohmeyer, M. and Teller, S., 1991. "Stabbing Isothetic Boxes and Rectangles in $O(n \log n)$ Time," UC Berkeley TR 91/634.

18. Horaud, R., 1987, "New Methods for Matching 3-D Objects with Single Perspective Views," *IEEE Trans. Pattern Anal. Machine Intell.,* **9**(3): 401–412.

19. Hu M.K., 1962. "Visual pattern recognition by moment invariants," *IRE Transactions on Information Theory,* IT-8: 169–187.

20. D. Huttenlocher, K. Kedem, and J. Kleinberg, 1992, "On Dynamic Voronoi Diagrams and the Minimum Hausdorff Distance for Point Sets Under Euclidean Motion in the Plane," *Proc. of the Eighth ACM Symposium on Computational Geometry.*

21. Huttenlocher, D.P., and Ullman, S., 1990 "Recognizing Solid Objects by Alignment with an Image," *Int. J. Computer Vision,* **5**(2): 195–212.

22. D. Jacobs and R. Basri, 1996, "3-D Recognition with Regions," *Forthcoming Tech. Report.*

23. Marr, D. and Nishihara, H.K., 1978. "Representation and recognition of the spatial organization of three-dimensional shapes," *Proc. of the Royal Society, London,* **B200**: 269–294.

24. Megiddo, N., 1996. "Finding a Line of Sight Thru Boxes in d-Space in Linear Time," IBM Technical Report.

25. O. Menadeva, "The robustness to Occlusions in Recognizing Planar Objects with Regions," *The Weizmann Inst. of Science, Forthcoming M.Sc. Thesis.*

26. Pentland, A., 1987, "Recognition by Parts," *Proc. of the First Int. Conf. on Computer Vision*: 612–620.

27. Persoon E. and Fu K.S., 1977. "Shape descimination using Fourier descriptors," *IEEE Transactions on Systems, Man and Cybernetics* **7**: 534–541.

28. Reeves A.P., Prokop R.J., Andrews S.E., and Kuhl F.P., 1984. "Three-dimensional shape analysis using moments and Fourier descriptors," *Proc. of Int. Conf. on Pattern Recognition*: 447–450.

29. Richard C.W. and Hemami H., 1974. "Identification of three dimensional objects using Fourier descriptors of the boundry curve," *IEEE Transactions on Systems, Man and Cybernetics,* **4**(4): 371–378.

30. Ruckledge, W., 1995. "Locating Objects Using the Hausdorff Distance," *Fifth Int. Conf. on Comp. Vis*:457–464.

31. Sadjadi F.A. and Hall E.L., 1980. "Three-dimensional moment invariants," *IEEE Transactions on Pattern Analysis and Machine Intelligence,* **2**(2): 127–136.

32. M. Sharir and S. Toledo, 1994, "Extremal Polygon Containment Problems," *Comput. Geom. Theory Appls.* 4:99–118.
33. Ullman, S. and Basri, R., 1991. "Recognition by linear combinations of models," *IEEE Trans. on PAMI*, **13**(10): 992–1006.

Stably Placing Piecewise Smooth Objects

Chao-Kuei Hung and Doug Ierardi

Department of Computer Science
University of Southern California

Abstract. Given a piecewise smooth object, its stable poses consist of all the orientations into which other initial orientations of the object will eventually converge under dissipative forces. The capture region for each stable pose is the set of initial orientations converging to the stable pose in question.

We employ duality to solve these two related problems. Our approach produces non-trivial combinatorial bounds on the complexity of these problems as well as asymptotically efficient algorithms. It also allows us to remove the non-singularity constraints in Kriegman's previous work on the latter problem, and to enumerate the degenerate cases in a systematic way. Our analysis leads to a significant reduction in the algebraic complexity for objects consisting of quadratic surface patches cut by planes. The practical value of this approach is demonstrated by the implementation of an efficient approximation algorithm for this subclass of objects.

1   Introduction

Given a piecewise smooth object K, we study how it may be placed stably on a horizontal and flat surface. We consider two versions of this problem.

The static version seeks to determine the set of *stable poses* of K, or the local minima of its gravitational potential energy as a function of its orientation with respect to the direction of the gravity. Such information is useful for example, in a model-based vision system, where the complexity of the recognition of static objects can be reduced as a result of reducing the number of unknown parameters by two.

The more general version seeks to determine not only the stable poses of K, but also a *capture region* for each stable pose p, namely the set of all initial poses released from which K will eventually roll into p. Consider the last step of the path-planning algorithm for a robot where the robot hand is about to place an industrial part on a flat surface. Knowledge of the capture region for the desired object pose (which must be stable) helps the path-planning algorithm determine the required precision of this operation. Comparison among the capture regions associated with various stable poses also leads to a better decision if the choice of the object pose is an option for the algorithm.

Kriegman [Kri91] has taken the direct approach to analyzing the stable pose problem. A complete implementation using this method would have to include sixteen different cases. His implementation, using homotopy method in solving the systems of algebraic equations, demonstrate the feasibility of the algorithm by dealing with objects composed of natural quadratic surface patches cut by planes without vertices. * The computation time, however, is not very attractive even for a simple object such as the one shown in Figure 1. (This is our reconstruction of an example taken from [Kri91].) In extending the work to capture regions, Kriegman [Kri95] gives a quadratic time algorithm for polyhedral objects. His direct approach using stratified Morse theory in analyzing the capture regions for an object composed of algebraic surface patches, however, has to make certain non-singularity assumptions that may be difficult to ensure in practice.

In this paper we take the dual approach to solve these two problems and extend Kriegman's results in several ways. We establish combinatorial upper bounds for the complexity of the potential energy function, from which the results on the stable poses and capture regions can be derived. This approach also allows us to remove the non-singularity constraints and to enumerate the degenerate cases in a systematic way. Furthermore, the algebraic overhead is greatly reduced when K consists solely of quadratic surface patches cut by planes.

2   Object Model and Assumptions

We assume that a piecewise smooth object K is described as an unordered set of N (open) faces $\{\varphi_1, \varphi_2, \ldots, \varphi_N\}$ whose union constitute a topological polyhedron embedded in E^3. [Cai68] Thus K is assumed to be compact but may have an arbitrary genus, need not be connected, and need not be a manifold. Furthermore, each 2-face is assumed to be a G^2 simply-connected surface patch, and each 1-face a G^2 curve segment. In order to give explicit formulas for the following results, we also assume that each face is a regular portion of an algebraic variety although this is not a requirement for the validity of the cell theorem. Given a face φ, we use *host* to refer to the entirety of the algebraic variety of which φ is a portion.

Finally, we assume that the coordinate system is so chosen that the origin coincides with the center of mass of K. We will denote the origin by $\mathcal{O}$.

3   Duality

The *pedal* of a surface is the loci of the perpendicular foot dropped from $\mathcal{O}$ to a tangent plane to the surface as the point of tangency moves on the surface. [BG92] We naturally define ped φ, the pedal of a 2-face φ to be the portion of its host's pedal which corresponds to taking the tangent planes only within the extent of φ. We extend this notion and define the pedal of a 1-face φ to be the collection of the perpendicular feet from $\mathcal{O}$ to the planes which have a contact

---

* Cases involving vertices are explicitly left out from the enumeration in his paper. They do not seem to pose any more theoretical or practical difficulties than the enumerated cases since the algebra for the former is simpler.

of order at least one with φ. Finally we define the pedal of a vertex to be the collection of such perpendicular feet to the planes incident on the vertex.

The inversion of a point $\mathbf{p}$ in space is the the point $\mathbf{p}/\|\mathbf{p}\|^2$. We define the dual of a face φ to be the point-wise inverted image of its pedal and denote it by $\tilde{\varphi} = \mathbf{inv\,ped}\,\varphi$. It can be shown that

Theorem 1 *In the arrangement of the duals of K's faces, the cell which contains the origin is compact and convex, and it is the polar set of K.*

The polar set $K^* \stackrel{\mathrm{def}}{=} \{\mathbf{x} \;:\; \mathbf{x} \cdot \mathbf{y} \le 1\}$ of a set K is a classical form of duality. [Lay92]

For objects consisting of algebraic 2-faces and edges, the dual faces can be computed by performing variable elimination. [Col75, IK93, Man93] We shall see, however, that this problem reduces to linear algebra on small matrices (4 by 4) when K consists of quadratic surface patches cut by planes, thus making the dual approach very attractive in practical terms.

4 The Potential Energy Function

Consider representing the boundary of the polar set K^* in spherical coordinates. The radius is thus a single-valued function of the longitude and latitude since K^* is convex. The graph of the radius function of the polar set is precisely the reciprocal of the potential energy of the object as a function of its pose. Several questions regarding the stable poses and their respective capture regions can be answered once we construct a description of this function. In particular, the stable poses correspond to the local maxima of this function. It is shown in [Hun95] that

Theorem 2 *Let K be a piecewise smooth object with N faces such that each face has a bounded number of subfaces. Then the boundary of K^* can be constructed in randomized expected time $O(N^{2+\epsilon})$ for arbitrary but fixed $\epsilon > 0$.*

The proof is by way of reduction to Sharir's algorithm [Sha93] for constructing the lower envelope in an arrangement of surface patches. Thus we have the following

Corollary 1 *There are at most $K^{2+\epsilon}$ stable poses for the object in Theorem 2. It can be computed in the same amount of time, barring algebraic overhead.*

Corollary 2 *The capture regions of the object in Theorem 2 have a structural complexity of $O(N^{4+\epsilon})$ and can be computed in deterministic time $O(N^{4+\epsilon})$.*

Moreover, it is possible to treat in a systematic way the degenerate cases and coincidences, which are hard even to enumerate using the direct approach. Consider three identical fat ellipsoids stacked on their "flat" sides as in Figure 2. Aside from the two stable poses corresponding to two one-point contacts (top and bottom), it is not hard to see that there are two other stable poses corresponding to the two contacts where each ellipsoid contribute a point and where the three points of contact line up. It is not obvious whether the latter stable poses should be considered one-point, two-point, or three-point contacts. When one looks at its potential energy function in the dual space, one realizes that in the latter case each local maxima is the intersection of three saddles, one of which can be considered gratuitous in forming the local maximum (Figure 3). In general such special cases can be resolved in the algorithm by examining only the local structure of the radius function as opposed to having to consider features that are far apart in the primal space. Distinguishing the local maxima among all "critical poses" where certain partial derivatives vanish is similarly facilitated by working in the dual space.

5 Quadratic Surface Patches Cut by Planes

The results in the previous section are more of theoretical interest since we have focused only on the combinatorial complexity. Kriegman's experimental results, however, show that the dominating factor in computing the stable poses is the algebraic overhead. Taking the direct approach, his implementation has to solve a system of quadratic functions in up to nine variables for simple parts made of cylinder sections and spheres. Fortunately the dual approach also reduces the algebraic complexity dramatically in the case of quadratic surface patches cut by planes. The variable elimination in the dualizing step reduces to simple linear algebra operations.

In the special case of quadratic surfaces, the inverted pedal becomes the classical duality in projective geometry with a slight modification (in fact just reflect every vector with respect to the origin). Thus to compute the dual of a quadratic face φ, we simply represent the host surface by a 4 by 4 real symmetric matrix, say Q and then invert Q — provided that the surface is non-degenerate and hence Q is non-singular. The boundary of the dual surface can also be obtained by mere linear algebra without going through the elimination process. Let $\mathbf{y}$ be the 4-vector representation of a cutting plane H of φ. The pole of H with respect to the host of φ dualizes to the cutting plane of $\tilde{\varphi}$ which determines the extent of the dual face, and is simply $\mathbf{y}Q^{-1}$.

We also have to give a recipe for computing the duals in the degenerate cases such as cones and cylinders. In this case we first compute $Q = U\Lambda U^T$ where U is orthogonal and Λ is diagonal. Let Λ' be the same as Λ except with the vanishing diagonal entry replaced by 1. Then $\tilde{\varphi}$ is the intersection of $U(\Lambda')^{-1}U^T$ with the plane whose coefficients are the eigenvector of Q corresponding to the vanishing eigenvalue.

Finally the duals of the edges can also be computed efficiently. Let γ be an edge whose host is the intersection of a non-degenerate quadratic hypersurface represented by Q with a plane represented by the 4 by 1 matrix N. We first perform a QR factorization and write

$$N = (U_1 U_2) \begin{pmatrix} R \\ 0 \end{pmatrix}.$$

where U_1 is 4 by 1, U_2 is 4 by 3, both orthogonal, and where R is upper-triangular and non-singular. Next write Q as

$$Q = (U_1 U_2) \begin{pmatrix} A & B \\ B^T & C \end{pmatrix} \begin{pmatrix} U_1^T \\ U_2^T \end{pmatrix} .$$

Then it can be shown that $\tilde{\gamma}$ is the (degenerate) quadratic surface $U_2 C^{-1} U_2$.

Putting all this together, we conclude that the dominant algebraic calculations involved in finding the stable poses of such objects are:

- Solving a system of 3 quadratic equations for each vertex of $\partial(K^*)$.
- Maximizing a quadratic function over the intersection of two quadratic surfaces for each edge of $\partial(K^*)$. Note that the intersection has a closed form parameterization.
- Maximizing a quadratic function over a quadratic surface for each dual face. This can be reduced to finding the root of a univariate polynomial of degree 6.

Note that this does not take into account the reduction to the lower envelope problem and its solution, which would incur additional algebraic overhead. In the next section we propose an approximation algorithm that may not be asymptotically optimal, but is extremely efficient in practical terms, as demonstrated by our implementation.

6   An Approximation Algorithm

In the dual space, consider shooting rays from the origin with nearly regular spacing between the rays, for example, doing so uniformly at the grid points of the longitudinal and latitudinal lines on the unit sphere. Create a planar graph whose vertices correspond to the rays and whose edges connect the neighboring rays in a natural way (such as the longitudinal sections and latitudinal sections in our example). For each ray find the first dual surface it hits and compute the tangent plane at the point of intersection. The planar graph can be refined by adding vertices between adjacent vertices which fall on different dual faces, adding edges between the newly added vertices which surround the same face of the graph and fall on the same dual edge, and finally adding new vertices which correspond to the intersections of the new edges. In this way we create a finer description of the potential energy function than is necessary. The combinatorial complexity is raised by a quadratic factor of the resolution of the rays (and reduced by a linear factor in the number of the object features) whereas the algebraic overhead required to construct it is substantially lower. Our partial implementation using C++ shows that the performance is rather realistic.

It takes roughly one minute to create, among other things, a mesh of about 100 by 50 (which is sufficient for this particular example) for the radius function of the polar set of a 6-piece pipe like Figure 1 (reconstructed from [Kri91]) even on an i486-33 running OS/2 or Linux. The radius function of the polar set is shown in Figure 4.

Such uniform sampling scheme in the dual space can be understood as weighing the sampling density (number of rays per unit area on a face), *in the primal space*, by the Gaussian curvature of the face since the latter is the limiting ratio between the area of the Gaussian sphere and the surface area in the primal space. Intuitively, we are sampling more frequently at badly curved places, and saving our "sampling bandwidth" at relatively area of the convex hull of the original object.

Acknowledgement

We are indebted to David Kriegman, who pointed out the applicability of our previous results on convex hulls to the stable pose problem and provided comments and stimulating discussions. We would also like to thank Francis Bonahon, Aristides Requicha, and Ken Goldberg for their valuable comments and references.

References

[BG92]   J. W. Bruce and P. J. Giblin. *Curves and Singularities*. Cambridge University Press, 2nd edition, 1992.

[Cai68] Stewart Scott Cairns. *Introductory Topology*. Ronald Press Company, 1968.

[Col75] G. E. Collins. Quantifier elimination for real closed fields by cylindrical algebraic decomposition. In *Second GI Conference on Automata Theory and Formal Languages*, volume 33, pages 134–183. Srpinger-Verlag, 1975.

[Hun95] Chao-Kuei Hung. *Convex Hull of Surface Patches: Construction and Applications*. PhD thesis, Computer Science Department, University of Southern California, 1995.

[IK93]   Doug Ierardi and Dexter Kozen. Parallel resultant computation. In John H. Reif, editor, *Synthesis of Parallel Algorithms*. Morgan Kaufmann, 1993.

[Kri91] David J. Kriegman. Computing stable poses of piecewise smooth objects. In *Computer Vision Graphics Image Processing: Image Understanding*, 1991.

[Kri95] David J. Kriegman. Let them fall where they may: Capture regions of curved objects and polyhedra. Technical Report 9508, Yale University, June 1995. Submitted to the International Journal of Robotics Research.

[Lay92] Steven R. Lay. *Convex Sets and Their Applications*. John Wiley & Sons, Inc., 1992. original edition 1982.

[Man93] Dinesh Manocha. Solving polynomial systems for curves, surface, and solid modeling. In *Proc. of ACM/SIGGRAPH*, 1993.

[Sha93] Micha Sharir. Almost tight upper bounds for lower envelopes in higher dimensions. In *Symposium on Foundations of Computer Science*, pages 498–507, 1993.

[SS85]   Otto Schreier and Emanuel Sperner. *Projective Geometry of n Dimensions*. Chelsea Publishing Company, New York, 1985.

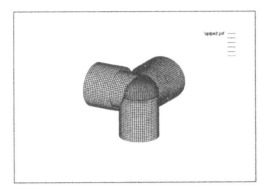

Fig. 1. A PVC pipe fitting. The center of mass is just outside the sphere in the octant wedged by the three branches.

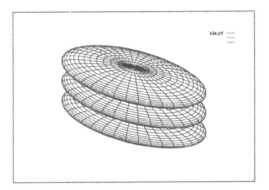

Fig. 2. Is this a one-point, two-point, or three-point contact?

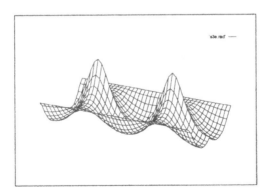

Fig. 3. The radius function of the stacked ellipsoids.

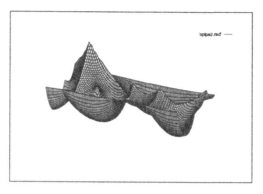

Fig. 4. The radius function of the polar set of the pipe fitting.

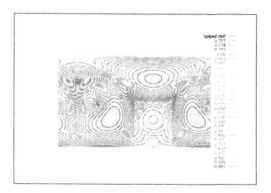

Fig. 5. The contour plot of the radius function of the polar set of the pipe fitting.

A Beam-Tracing Algorithm for Prediction of Indoor Radio Propagation

Steven Fortune

AT&T Bell Laboratories
Murray Hill, NJ 07974
sjf@research.att.com

Abstract. We describe a beam tracing algorithm that simulates radio propagation inside a building. With a triangulation-based spatial data structure, the algorithm is fast enough to provide propagation simulations for large buildings in a few minutes of computing time.

1 Introduction

Wireless communication networks using radio-frequency transceivers are important for many indoor applications. The design of such networks requires accurate prediction of radio propagation inside a building.

Radio propagation can be modeled fairly simply, using geometric optics. A building is a list of walls, each a polygon labeled with construction material. Radio propagates along a straight line, with power diminishing according to an inverse-square law. At a wall, part of the radio power passes through and part is reflected specularly, with proportions depending upon material and incidence angle. Diffraction is ignored in this model, as well as scattering at a wall or because of furniture and other clutter. To determine coverage, power is sampled at the lattice points of a regular grid.

Naive simulation of propagation is computationally challenging. In a large building there may be thousands of possible propagation paths between base transmitter and sample point, though only a few contribute significant power, and received power may need to be simulated at thousands of different sample points. The dynamic range of the simulation—the ratio between minimum and maximum received power—is 10^6 or more. The simulation is most useful if it is accurate over the entire dynamic range.

This paper describes a beam tracing algorithm that can be used to predict radio propagation. Each wall can be viewed as a partially reflecting mirror that illuminates a polyhedral cone, using reflected power from the base. If a sample point lies in an illumination cone, then there is a propagation path to the sample point that reflects off the corresponding wall; the actual path power may be reduced by the shadowing of intermediate walls. All propagation paths, up to some fixed number of reflections, can be found with such illumination cones.

The algorithm uses a spatial data structure based on triangulations. This data structure answers ray-shooting queries in constant time (as measured experimentally) and allows efficient detection of all environment obstacles that lie in a polyhedral query cone. Implementation experiments are reported below.

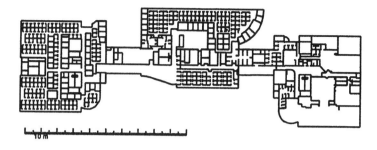

Fig. 1. Plan view of AT&T Mt. Airy first floor; this model has about 2315 walls.

This paper does not address the accuracy of the propagation model itself. Many factors might affect prediction accuracy: the presence of furniture, inaccuracies in the building model, the modeling of reflection and transmission losses, other radio effects such as scattering or diffraction, etc. The literature contains some comparisons between predictions and measurements, for this and similar models [4, 10].

The algorithm in this paper have been incorporated into a wireless system engineering tool called WISE [1]. WISE includes a powerful graphical user interface that allows system engineers to choose base placements interactively, visualize propagation predictions, and optimize base placement automatically.

Other work. Beam tracing is a standard technique in computer graphics rendering [11]; its use for radio propagation has been independently proposed by Takahashi *et al* [9]. Propagation can also be predicted using ray tracing, i.e. by following paths in a large but discrete set of propagation directions[4, 5, 7, 10]. Beam tracing is competitive with ray tracing in performance and has fewer sampling artifacts; the full version of this paper [2] gives a detailed comparison.

Other practical spatial data structures can be used to answer ray-shooting and related queries. Held *et al* [3] decompose space into tetrahedra; Rajkumar *et al* [7] use binary space partition trees.

2   Radio background

The books by Parsons[6] and by Ramo *et al*[8] are general references for radio propagation.

The environment consists of a set of *walls*, each of which is a polygon with a small number of edges, e.g., a rectangle. Each wall has a type that indicates its construction material and hence radio properties. A typical environment might have between 100 and 10000 walls (figure 1).

A *propagation path* is a directed polygonal chain such that at each internal vertex, the path is generated by reflection off a wall using Snell's law, that is, the direction along the path after the vertex is the reflection in the plane of the wall of the direction before the vertex. A propagation path may pass through a wall.

A propagation path p has an associated *path loss*

$$l_p = \frac{\prod_i \rho_i \prod_j \tau_j}{|p|^2} \tag{1}$$

where $|p|$ is the length of the path in meters; ρ_i is the ith *reflection coefficient*, i.e., loss due to reflection; and τ_j is the jth *transmission coefficient*, i.e., loss due to a wall intersection that is not a reflection. Measured in dB, path loss is $10 \log_{10} l_p$.

The (free-space) received power at one meter is given by

$$\beta_{1m} = P \left(\frac{\lambda}{4\pi} \right)^2$$

where P is the power input to the transmitter and λ is the wavelength in meters. The *received power* at a sample point is

$$\beta = \beta_{1m} \sum_p l_p \tag{2}$$

where the sum runs over all propagation paths from base transmitter to sample point. Measured in dBm, received power is $10 \log_{10} \beta$, assuming transmitter power is measured in milliwatts (mw).

Typical transmitter power is 34 mw; at 2 Ghz this yields a free-space received power of about -23 dBm at one meter. A received power of -70 dBm is typically required for reliable communication, implying a free-space coverage radius of about 220 meters. It is appropriate to model paths with even lower power, say down to -90 dBm, since several paths could add together to meet the -70 dBm threshold, and since it is desirable to model interference between transmitters operating on the same frequency.

A *coverage calculation* determines received power at a set of sample points. Typically there are infinitely many propagation paths to a sample point. Hence a coverage calculation is usually an approximation, with propagation paths restricted to those of relatively high power or with a limited number of reflections.

Theoretical transmission and reflection coefficients can be determined using a multilayer dielectric model [8] derived from estimated properties of construction materials. For example, a sheetrock wall might be modeled as a half inch of plaster, 3.5 inches of air, and a half inch of plaster. For a fixed wall material and wall orientation, the coefficients depend upon the angle of incidence of path to wall and can be determined to adequate accuracy with table lookup indexed

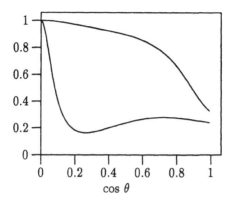

$\cos \theta$

Fig. 2. Theoretical reflection coefficient as a function of incident angle: sheetrock (top) and wood over concrete (bottom). Transmission coefficient is 1 minus reflection coefficient.

by discretized angle. Figure 2 plots the coefficients for sheetrock and 'wood over concrete' as a function of incident angle; both walls are vertical. For most walls, the transmission coefficient is maximized if the path is perpendicular to the wall; the reflection coefficient approaches 1 if the path is almost parallel to the wall. In practice, theoretical reflection and transmission coefficients may need to be altered to model losses due to scattering, construction inhomogeneities, furniture, etc.

3   The beam tracing algorithm

Fix the position of the base transmitter. The *reflection cone* $C(w)$ for a wall w consists of all points reachable by a single-reflection path off wall w. Let v be the reflected image of the base in the plane of the wall; v is a *virtual transmitter*. It is immediate that a sample point is reachable by a single reflection off wall w exactly if the line segment from v to the sample point intersects w at the reflecting point. Thus the reflection cone is a truncated polyhedral cone. It has apex v; each plane through v and one of the edges of w bounds the cone; the cone is truncated by the plane of w, deleting the portion of the cone from v to w. See figure 3.

In general, for a sequence $w_1, \ldots, w_k$ of walls, the *reflection cone* $C(w_1, \ldots, w_k)$ is the set of all points reachable by paths that reflect off walls $w_1, \ldots, w_k$ in order. It has apex the *iterated virtual transmitter* $\hat{v}$, obtained by reflecting the transmitter successively in the planes of $w_1, \ldots, w_k$. $C(w_1, \ldots, w_k)$ is bounded by each plane through $\hat{v}$ and an edge of the polygon $w_k \cap C(w_1, \ldots, w_{k-1})$. As before, $C(w_1, \ldots, w_k)$ is truncated by the plane of w_k.

The beam tracing algorithm assumes an *a priori* bound on the number r of reflections. It can be given simply as

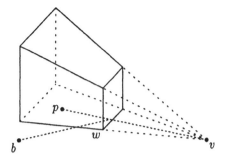

Fig. 3. Reflection cone (outlined in solid lines). b is base, v is virtual transmitter. Reflecting path from b to p intersects wall w at same point as line segment vp.

```
for each reflection cone C
    for each sample point p inside C
        compute power of path to p.
```

The reflection cones can be generated by a simple recursive process: there is a one-reflection cone $C(w)$ for each wall w, and a cone $C(w_1, \ldots, w_k)$ for each wall intersecting $C(w_1, \ldots, w_{k-1})$. Section 4 describes an efficient method that enumerates the walls inside a reflection cone.

In the worst case, the number of nonempty reflection cones grow as n^r, where n is the number of walls. More typically a reflection cone intersects about $n^{1/2}$ walls, so the number of cones grows as $n^{(r+1)/2}$. See section 5 for experimental data.

Given sample point p lying in a reflection cone, it is easy to find the corresponding propagation path to p. The power computation using formulas (1) and (2) requires the discovery of the walls that intersect the propagation path. Section 4 describes a spatial data structure that can be used to trace a propagation path, finding the intersected walls in time that is approximately constant per wall.

Even with the spatial data structure, path tracing is the dominant cost of the algorithm. Consequently, it is useful to avoid a path power computation if it will not contribute significant power to a sample point; here significant power is defined as a fraction, say .01, of the power already accumulated at the sample point.

Let w be the last wall in the reflection sequence of a reflection cone C, and let m be an interior point of $C \cap W$, say the centroid of the vertices of $C \cap W$. Cone C is assigned an *estimated power* β_C, defined as the power from the base to m (including the final reflection coefficient off wall w).

The value β_C can be used to filter path power computations, in two ways. First, if β_C is less than an absolute threshold, say -90 dBm, then C is ignored (as are any reflection cones produced by walls intersecting C). Second, suppose

β_C is above threshold and sample point p lies inside C. An upper bound estimate of the path power to p is $\beta_C(d_m/d_p)^2$, where d_m and d_p are propagation path distances to m and p, respectively. If this upper bound cannot add significant power to p, then the path is ignored. Otherwise the path is traced and path power computed; notice that β_C does not affect the actual path power computation.

It is possible that a filtered-out path in fact contributes significant power to a sample point. This can happen if there is a wide variation on the path power to points of $C \cap w$, and β_C happens to be a low estimate. The probability of such erroneously-filtered paths is minimized by restricting filtering to narrow cones, where a cone C is *narrow* if the maximum difference between any two direction vectors from the virtual transmitter of C to points in C is small, say ten degrees.

The full paper [2] discusses an algorithm that sweeps through each reflection cone, maintaining the 'shadow partition' induced by wall transmission losses. This algorithm eliminates the need to trace each propagation path to a sample point. However it is more expensive to maintain the shadow partition (roughly n^3 per reflection cone) than to do the path tracing.

4 Data structures

We describe a triangulation-based spatial data structure that can be used to answer two geometric queries: report the walls intersected by a query line segment, and report all walls inside a convex polyhedral query cone. To use the data structure, each wall polygon must either be horizontal, of arbitrary shape, or a vertical rectangle, i.e., with two sides parallel to the z-axis and top and bottom horizontal. Most building walls can be modeled with polygons of this type, though e.g. cathedral ceilings cannot.

The triangulation data structure. A *distinguished z-value* is a z-coordinate of a wall vertex. The data structure consists of a set of triangulations. For each distinguished z-value, there is a triangulation of the edges of the horizontal walls at that z-value. Any triangle interior to a horizontal wall is marked with the wall that contains it. For each open interval between two adjacent distinguished z-values, there is a triangulation of the horizontal cross-section; notice that the horizontal cross-section consists of an edge for each vertical wall spanning the z-interval, and that the cross-section is constant over the interval. Each edge of the triangulation is marked with the wall that contains it, if any.

The space required to store the triangulation data structure is linear in the number of horizontal walls plus $\sum n_i$, where n_i is the number of vertical walls that span the ith z-interval. In the worst case $\sum n_i$ is quadratic in the total number of vertical walls. However, most building models have only a few distinguished z-values per building floor, perhaps one for the floor, one for the false ceiling, and one for the true ceiling; and most walls span at most a single floor. Hence in practice, the sum is linear in the number of walls.

Line segment queries. To report the walls intersected by a query line segment, first split the line segment into subsegments so that each subsegment lies within

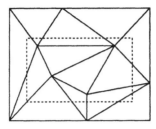

Fig. 4. Query polygon (dashed) separates triangulation into inside and outside; graph search can report triangles and edges inside.

a single z interval. The vertical walls intersected by a single subsegment can be determined using a walk in the corresponding triangulation, using the xy-projection of the subsegment. The horizontal wall crossed at each distinguished z-value, if any, can be discovered by a point location in the corresponding triangulation.

Polyhedral cone queries. The following strategy reports all walls intersected by a convex polyhedral query cone; it is repeated for each distinguished z-value or interval between z-values.

The query cone intersects the horizontal plane through a distinguished z-value in a convex polygon. The algorithm walks each edge of the polygon in turn, marking the triangulation edges intersected by polygon edges. See figure 4. The edges partially or fully inside the polygon can then be enumerated by a standard graph search, starting with the marked edges. Each triangle incident to an enumerated edge is examined to see if it is marked with a horizontal wall.

The approach is similar for an interval of z-values. Consider the subset of the query cone that lies within the z-interval. The orthogonal projection onto the xy-plane of this subset is a convex polygon. The edges of this polygon are used for the walk, as before. In this case, the graph search examines edges of the triangulation for intersected walls, rather than triangles.

Performance. For a line segment query, the number of triangulation edges traversed per reported wall varied between 2.5 and 6, depending upon the uniformity of building walls. For the polyhedral cone query, the number of triangulation edges visited per reported wall usually varied between 6 and 20, though sometimes it was more than 100.

For point location queries, a rectangular grid was superimposed on the triangulation; each grid cell stored the triangle containing the center of the cell. A query point can be located by a triangulation walk from the stored triangle of the cell containing the query point. On average, the walk traversed between .5 and 2.2 triangulation edges, indicating that this simple grid is quite adequate.

Building	Floors	Walls	size (m)	Sample points
Crawford	1	81	15 × 130	418
Hynes	4	1436	130 × 170	5240
Indiana	2	278	120 × 250	7060
Middletown	1	1367	200 × 200	3651
Mt Airy	3	2316	60 × 220	2449
Prudential	2	1175	65 × 85	1366

Fig. 5. Sample buildings.

Building	reflections				
	1	2	3	4	5
Crawford	:01	:04	:10	:27	:51
Hynes	:12	:53	2:39	6:18	14:08
Indiana	:11	:37	1:44	4:31	10:24
Middletown	:14	1:09	4:04	13:09	39:20
Mt Airy	:11	1:40	8:03	32:52	103:03
Prudential	:07	:44	4:25	19:46	66:31

Fig. 6. Running time of beam tracing, mm:ss.

5 Experimental results

The beam-tracing algorithm was implemented in C++ using the geometric data structures described above. All times are for an SGI R4400, running at a clock-doubled rate of 100 Mhz (for an effective clock rate of 200 Mhz).

Figure 5 gives a capsule description of the buildings used in the experiments. For each building, a representative location of the base was chosen and used throughout. The sample points were the lattice points of a regular grid with 2 meter spacing on a horizontal plane inside the building. For all runs, the frequency was 1.92 Ghz and the transmitter power was 34 mw.

The beam tracing algorithm was run with the filtering described in section 3, and in some cases without the filtering. For the filtered runs, a reflection cone was ignored if its estimated power was −90 dBm or below; a propagation path to a sample point was ignored if it was estimated to contribute less than .01 times the power already accumulated at the sample point. Figure 6 gives the running times of the filtered beam tracing algorithm. For one or two reflections, more than 50% of running time is spent doing triangulation walks; for higher reflections, the manipulation of reflection cones is dominant. The unfiltered running times are not given but are much worse, up to 30 times slower in some cases.

The average difference between filtered and unfiltered runs was typically .1 dB for a one reflection map, increasing to between .4 and .7 dB at three reflections;

Building	refl	cones ($\times 10^3$)	cones > −90dBm ($\times 10^3$)	paths ($\times 10^6$)	walks ($\times 10^3$)	walks/ sample	walks/ cone
Crawford	2	3	.6	.2	19	47	31.0
Crawford	3	61	2.2	3.4	32	78	14.2
Crawford	4	1269	4.5	57.2	41	98	9.1
Mt Airy	2	131	19.1	1.2	230	102	12.0
Mt Airy	3	4873	110.6	23.1	614	273	5.5
Mt Airy	4	155212	407.8	436.5	1213	539	3.0
Prudential	2	54	18.3	.7	114	84	6.2
Prudential	3	1434	105.9	11.9	276	202	2.6
Prudential	4	30886	363.8	189.9	537	393	1.5

Fig. 7. Beam tracing: reflection cones, total and with estimated power above threshold; number of propagation paths; number of path power computations ('walks') performed; and path power computations per sample point and per cone.

the standard deviation varied between .2 and .4 dB. The maximum difference was 2 to 5 dB, with larger values at higher reflections.

Figure 7 gives more details on selected runs. The number of reflection cones roughly fits the prediction of $n^{(r+1)/2}$, while the number of cones with estimated power above the −90 dBm threshold is much smaller. The column labeled 'paths' gives the number of propagation paths, i.e., the number of times a sample point lies inside a reflection cone (ignoring the threshold). The column labeled 'walks' gives the actual number of path power computations performed; clearly, the filtering has dramatically reduced this number from the number of propagation paths. The number of path computations per sample point gives an absolute measure of the efficacy of filtering.

6   Conclusions

The beam tracing algorithm is many orders of magnitude faster than a naive algorithm [10]; useful predictions can be obtained within a few minutes. The speedup is obtained by exploiting the geometry of the problem and by utilizing appropriate geometric data structures.

Similar techniques may be applicable for prediction in outdoor urban or suburban environments. The radio theory is somewhat similar, though diffraction becomes an important propagation mechanism. Computationally, the prediction problem is somewhat different as walls are modeled as opaque, with reflections only.

7 Acknowledgements

Thanks to Brian Kernighan, Margaret Wright, and Reinaldo Valenzuela for helpful comments.

References

1. S. J. Fortune, D. M. Gay, B. W. Kernighan, O. Landron, R. A. Valenzuela, M. H. Wright, WISE design of indoor wireless systems: practical computation and optimization. *IEEE Computational Science and Engineering*, Spring 1995, pp. 58–68.
2. S. J. Fortune, Algorithms for prediction of indoor radio propagation, manuscript, 1995.
3. M. Held, J. Klosowski, J.S.B. Mitchell, Evaluation of collision detection methods for virtual reality fly-throughs, *Proceedings 7th Canadian Conference on Computational Geometry*, 1995, pp. 205–210.
4. P. Kreuzgruber, T. Bründl, W. Kuran, R. Gahleitner, Prediction of indoor radio propagation with the ray splitting model including edge diffraction and rough surfaces, *Proceedings of the 1994 IEEE 44th Vehicular Technology Conference*, IEEE, 1994, 878–882.
5. J.W. McKown, R.L. Hamilton, Jr. Ray tracing as design tool for radio networks, *IEEE Network Magazine*, Vol 5, pp. 27–30, 1991.
6. J.D. Parsons, *The Mobile Radio Propagation Channel*, John Wiley and sons, 1992.
7. A. Rajkumar, B. Naylor, F. Fiesullin, L. Rogers, Predicting RF coverage in large environments using ray-beam tracing and partitioning tree represented geometry, Technical Memorandum, AT&T Bell Laboratories, 1995.
8. S. Ramo, J.R. Whinnery, T. Van Duzer, *Fields and Waves in Communication Electronics*, 3rd edition, John Wiley and Sons, 1993.
9. S. Takahashi, K. Ishida, H. Yoshiura, A. Nakagoshi, An evaluation point culling algorithm for radio propagation simulation based on the imaging method, *Virginia Tech Fifth Symposium on Wireless Personal Communications*, pp. 13.1 – 13.11, 1995.
10. R. A. Valenzuela, A ray-tracing approach to predicting indoor wireless transmission, *Proc. 1993 IEEE 43rd Vehicular Technology Conference*, IEEE Press, Piscataway, NJ, 1993, pp. 214–218.
11. A. Watt, M. Watt, *Advanced Animation and Rendering Techniques: Theory and Practice*, Addison-Wesley, 1992.

Extracting Geometric Information
from Architectural Drawings

Brian W. Kernighan[1] and Christopher J. Van Wyk[2]

[1] Bell Laboratories, Murray Hill, NJ 07974, USA,
bwk@research.bell-labs.com
[2] Department of Mathematics and Computer Science, Drew University,
Madison, NJ 07940, USA, cvanwyk@drew.edu

Abstract. We describe some experiments on a geometric problem encountered while developing a tool for predicting the behavior of indoor wireless communications systems. The problem is to extract information about the walls of a building from a machine-readable description. These descriptions contain much irrelevant or redundant information; the task is to eliminate the former and to reduce the amount of the latter. A key geometric idea is to use plane sweep to find line segments in the drawing that belong to the same wall. A great gulf lies between this simple idea and its realization, however, and any solution must involve a tradeoff among factors like accuracy, speed, and the degree and nature of human intervention.

1 Introduction

For the past two years one of the authors has been part of a group developing the WISE (Wireless System Engineering) system for design and optimization of indoor wireless communications systems. Cellular and cordless phones, wireless networks, active badges, and radio-controlled shelf tags in supermarkets are among the many applications of this kind of system. WISE requires as input a description of the coordinates and composition of the walls of a building, and myriad parameters, including power, frequency, antenna types, and signal to noise ratios. A family of programs computes radio energy throughout the building, to predict the behavior of the system, to minimize system cost by varying the locations of transceivers, and to analyze coverage, sensitivity, and other properties. The overall system is described in [FGK+95].

Developing WISE required solutions to a variety of computational geometry problems. Techniques that reduce computational time from infeasible to practical are described in [F96]. This paper is about a smaller, subsidiary problem that illustrates some pragmatic issues that lie between the important ideas of computational geometry as viewed by the computer science community and the difficult real-world problems posed by potential users of those ideas.

The problem is to determine the coordinates and composition of the walls of the building from whatever representation is given. Although easy to state and to understand, this seems to be the most difficult problem faced by users of WISE. Recent buildings, say those built within the past decade or two, are often described in some computer-aided design system, usually AutoCAD "DXF" files [MvR94]. We attacked the problem of processing AutoCAD drawings mechanically to make them more suitable as building representations for WISE.

2   Decoding AutoCAD DXF Files

A DXF file contains a header section that defines parameters, a blocks section that defines drafting symbols and common constructs like door frames, and an entities section that contains all the items in the drawing. Each entity has a type, coordinates, color, and other parameters, and is associated with a "layer" of the diagram; the color and layer may indicate its role in the diagram or building.

Since DXF files are in ASCII, we can use standard Unix tools to manipulate them. Thus, a 150-line AWK [AKW88] program parses DXF files and outputs an ASCII representation of line segments: each segment is represented by the (x, y) coordinates of its endpoints, an assumed z-value, and any information about composition that can be guessed from the layer and color fields. This program also does any necessary coordinate transformations, like clipping, inverting y, and converting feet to meters. Other AWK programs process these line segments; the Unix sort command puts them into various useful orders; and Tcl/Tk [O94] lets us quickly write programs to view the data interactively before, during, and after processing.

3   Removing Extraneous Features

Figure 1 shows the ground floor of AT&T's Learning Center, a 3-story conference center and hotel in Basking Ridge, NJ. Although some information (like wiring and pipes) had already been filtered out before we got the data, it is apparent from what remains that DXF files are meant to produce drawings for people. They contain features like closely spaced, short line segments that represent stairs; arcs or lines that indicate the sweep of doors; cross-hatched elevator shafts and building columns; dotted and dashed grid lines; line drawings of plumbing fixtures; and textual annotations. For instance, the lines at the top of Figure 1 underline text that titles the drawing, and the "arrowhead" marks the nominal North.

Thus, part of the processing is to weed out such extraneous features. After some experiments, we arrived at the following heuristics:

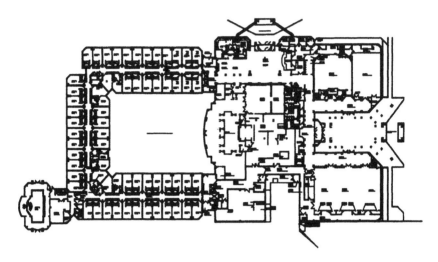

Fig. 1. AutoCAD drawing of AT&T Learning Center ground floor (~ 155m × 80 m)

- Discard all entities except LINEs and POLYLINEs.

- Eliminate all line segments that are too short to matter to the radio propagation model.

- Eliminate line segments that serve only to connect annotation text to a feature of the drawing; these often follow a TEXT entity directly and share an endpoint with it.

- Eliminate any line segment that is not drawn in CONTINUOUS mode; it is likely to be a grid line.

The first two heuristics are fairly general and likely to apply to many buildings. The second two, however, apply only to some of the buildings we have analyzed for WISE users, and might well need to be modified or augmented for other representations. Each new building — indeed, different floors of the same building — can add to a growing list of heuristics that have proven useful for removing extraneous features.

One lesson we learned from the exploration and experiments that led to these heuristics is that a DXF file is in no way a complete or accurate representation of a building. It is more realistic to regard the diagram as a comment: at best it bears some relation to some version of the building at some time, but it should not be confused with the real thing. (*Proof.* In Figure 2, only half the hotel rooms have toilets.)

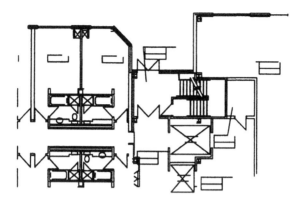

Fig. 2. Detail from Figure 1

4 Coalescing Nearby Walls

The close-up view in Figure 2 shows that the DXF file represents each wall as a collection of line segments that define the wall's perimeter in the plan view: each "logical wall" has at least two line segments, one for the room on each side of the wall. WISE, however, expects the building representation to contain "logical walls," since it models each such wall as a plane surface with zero thickness but specified dielectric properties appropriate to a real (physical) wall.

To use WISE on a building representation like Figure 1, we would need to modify the specified dielectric properties of each material to adjust for the "doubling" of most of the walls. The run-time for WISE on this problem would be at least four times the run-time of a representation with "logical walls." (The time complexity is $O(n^{b+1})$, where b is the number of bounces in the ray-tracing process and n is the number of walls in the building.) Thus, we seek to replace most of the multiple walls in the drawing by appropriate single logical walls.

Some sort of plane-sweep algorithm seems like a promising way to identify and coalesce nearby walls. One problem is to pin down exactly *what* sort of plane-sweep: we need to define what it means to be "nearby" and to "coalesce" two walls. Since we were not sure the output would be useful — significantly smaller, yet still detailed enough to capture important properties of the original building — we needed a quick-and-dirty implementation to let us explore various possibilities and see how good the results looked.

We developed our program in small pieces connected in a pipeline. The heart of the processing is performed by a 40-line AWK program that computes the transitive closure of the following relation on horizontal segments: segments S and T are "nearby" if the horizontal projections of S and T intersect and the vertical distance between S and T is at most h. An equivalence class under the transitive closure of "nearby" is a collection of line segments $s_1, \ldots, s_k$. Each such equivalence class is represented in the output by a line segment whose endpoints have x-coordinates equal to $\min\{s_i.xmin \mid 1 \le i \le k\}$ and $\max\{s_i.xmax \mid 1 \le i \le k\}$, and whose y-coordinate is $(\min_{1 \le i \le k} s_i.y + \max_{1 \le i \le k} s_i.y)/2$.

To compute the transitive closure of "nearby," the program sweeps over a sequence of horizontal line segments that have been sorted by y-coordinate. At each segment S, it begins to grow a rectangular box. Initially, the top and bottom of the box coincide with S. As the program considers segments that lie later in the sequence than S, whenever it finds a segment T that lies "nearby" the top of the box, it "coalesces" T into the box: the vertical edges of the box are set to coincide with the horizontal extrema of the box and of T, the top of the box is set at the y-coordinate of T, and T is removed from the sequence of segments.

This implementation of plane-sweep will win no awards for asymptotic efficiency, but it led us to discover some interesting aspects of the problem:

- Some input line segments are not quite horizontal; we added a tolerance to accommodate this.

- Some input line segments almost touch at their endpoints; we added another tolerance that accounts for this.

- The picture of the building is still recognizable when only the horizontal line segments output by the coalescing algorithm are drawn.

Emboldened by the third discovery, we extended the coalescing process to the three other principal compass directions: positive and negative unit slope, and vertical. The easiest way to do this was to extract walls of these slopes from the original data, rotate them so they were horizontal, pass them through the original coalescing program, then rotate them back into their original positions.

Figure 3 shows the same floor of the Learning Center building as Figure 1, but after it has been processed by our program. (For this test, the h of the "nearby" relation was taken to be ½ foot, walls that did not lie very close to one of the principal compass directions were preserved unchanged from input to output, and features shorter than 1½ feet on a side were eliminated after the coalescing operation had been performed.) The drawing is clearly similar but much simpler. Equally clearly, places remain where the process has not simplified things as much as we might have hoped, for example, the residual drafting lines at the top.

One (coarse) measure of the quality of the coalescing step is the reduction in the number of walls. Table 1 shows some computational results for parts of three buildings.

It is difficult to quantify how well the output from our program resembles the original building. For some buildings, however, we can compare the output to the results of a human abstracting "significant" features from DXF input. Figure 4 shows the same building as extracted by a person working for five to ten hours. There are 1572 walls in the manually extracted version and 1291 in the automatically extracted version.

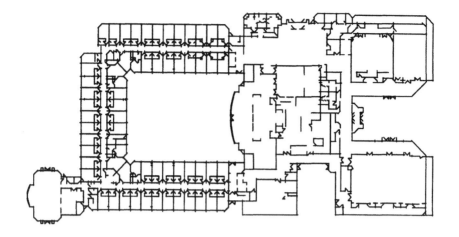

Fig. 3. Learning Center after wall coalescing

	Learning Center Floor 1	Middletown Floor 2	Red Hill Bldg 2 Floor 2
DXF source lines	691K	199K	89K
Total "walls"	24109	9782	3744
Less non-walls	9144	9782	3744
After coalescing	1994	3051	951
Horizontal	660	1153	430
Vertical	756	1192	388
Slope ±1	412	556	40
Other slanted	166	150	93
Less small features	1291	1657	487

Table 1. Some experimental results

The most important measure of the success of our program, of course, is how well it allows WISE to predict radio propagation. We have done limited experiments to compare propagation predictions on automatically and manually extracted drawings of the same building. To oversimplify somewhat, the general shape of coverage is similar, but until wall composition and ceiling location are set more carefully, detailed comparisons are inappropriate.

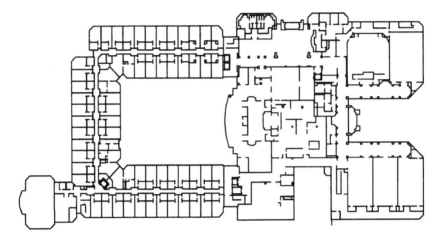

Fig. 4. Learning Center extracted by hand

5 Remaining Problems

For the buildings we have examined, wall composition is not indicated in the DXF file. For now, we use color information to eliminate some constructs that are clearly not walls, then make educated guesses about what remains. In the Learning Center, for example, structural walls appear to be denoted by two colors — one for exterior, another for interior. We have called the former concrete and the latter sheetrock, which is typical of such buildings.

None of the DXF files that we have seen include useful z information. To permit a sensible WISE simulation, we have adopted the expedient of using the bounding box to synthesize a floor, a ceiling, and a plenum one meter below the ceiling, again typical of most office buildings. Some kind of alpha-hull might be a better choice than the bounding box as a false floor or ceiling, but there is no point getting carried away by analysis: it is difficult to imagine any automatic method that would recognize the open-air courtyard in the Learning Center.

One of the drawbacks of the sweep algorithm is that the resulting horizontal and vertical walls need not meet at a corner. For example, the input in Figure 5(a) can produce any of the outputs of Figure 5(b), among others; only 4 of the 16 possible outcomes produce a closed corner. These gaps make only a small difference to predictions about radio propagation: there is leakage through the gaps, but the number of such paths is small, since the default grid upon which coverage is computed is about 2½ meters on a side. The imperfect corners are visible even in Figure 3, however, so some subsequent cleanup might be desirable for purely cosmetic reasons.

We were fortunate to find building drawings in DXF format. Given only paper blueprints, we would need to scan them, use image-analysis techniques to discover the walls, then "beautify" the results [PVW85] to join corners, align walls, and the like.

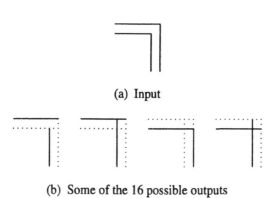

(a) Input

(b) Some of the 16 possible outputs

Fig. 5. Possible results of merging process

([RMM+94] describes an alternative, bitmap-based approach.) An alternative is to enter significant points from the drawing using a digitizing tablet; experience suggests that this can be done for a building of modest size in a few hours per floor. The techniques we have described could be used in a variety of ways to aid digitization: they might be used to produce a "first draft" for subsequent refinement by hand; or they might be used to highlight possible errors after the user has completed a digitization by hand. Because the DXF files exhibit such idiosyncratic properties, some human intervention is unavoidable, but there are ways to use these techniques to help it along.

6 Conclusions

Our experience taught us again the importance of an experimental approach. Real problems are rarely specified in a neat, precise form, so it is essential to be able to try things. Only through experiments could we establish which parts of the problem matter, what assumptions seem valid, what will (or won't) work in practice, and what might be worth doing. We proved no general theorems about our techniques, although it was occasionally useful to think about what was causing some would-be theorems to fail.

The evolution of our treatment of small features offers an example of the value of experimentation. At first we discarded small features early in processing, to reduce the amount of data to be considered by later phases. Later, we noticed that many small features appear at corners, where their presence during coalescing could help to make walls meet properly. This led us to make removal of small features the final processing step.

As an example of thinking about theorems, an earlier version of our program swept over the plane, coalescing nearby segments only when both lay within a narrow sweep-window. This made the output dependent on the sweep direction, so the program did not compute a relation at all. Our final program is slightly more complicated, but it computes a well-defined mathematical function of its input. The resulting output is also about 10% smaller and looks better.

For ease of experimentation, it was invaluable to have a simple uniform representation of information. Fortunately, DXF, though complicated and clumsy, is well documented and represented in ASCII. We converted it to an even simpler format: one line per entity, each with a type, a serial number, its coordinates, and its parameters. This straightforward data representation meant we could easily use existing tools to sort, select, summarize, transform coordinates, and the like. We believe that a set of small tools that can be connected in a pipeline proved more useful for our exploratory experiments than would a library of computational geometry code that required us to program in C or C++.

For geometrical applications in two or three dimensions, it is essential to be able to see the results of computations. We used Tcl/Tk [O94] extensively. Tcl/Tk was originally intended for creating graphical user interfaces for the X window system, but its "canvas widget" is an extremely flexible and convenient tool for visualizing line drawings. For instance, a 25-line Tcl/Tk program implements a viewer that lets us zoom in and out and pan over building diagrams, identify any object with a mouse click, and hide or color any class of objects in the picture.

Simple algorithms and data structures — sorting, files, and one-dimensional arrays — sufficed for our exploratory work. Should wall extraction turn out to be sufficiently attractive that it will be heavily used in practice, then we might need a proper implementation of an efficient, robust plane-sweep algorithm.

Acknowledgments

We are grateful to Orlando Landron and Chris Read for providing DXF data for our experiments, and to Jon Bentley for helpful comments on this paper.

References

[AKW88] A. V. Aho, B. W. Kernighan, and P. J. Weinberger, *The AWK Programming Language*, Addison-Wesley, 1988.

[FGK+95] S. J. Fortune, D. M. Gay, B. W. Kernighan, O. Landron, R. A. Valenzuela, and M. H. Wright, "WISE design of indoor wireless systems: practical computation and optimization," *IEEE Computational Science and Engineering* 2, Spring, 1995, 58–68.

[F96] S, J. Fortune, "A beam tracing algorithm for prediction of indoor radio propagation," these proceedings.

[MvR94] J. D. Murray and W. vanRyper, *Encyclopedia of Graphics File Formats*, O'Reilly, 1994.

[O94] J. K. Ousterhout, *Tcl and the Tk Toolkit*, Addison-Wesley, 1994.

[PVW85] T. Pavlidis and C. J. Van Wyk, "An automatic beautifier for drawings and illustrations," *Computer Graphics: Proceedings of SIGGRAPH '85* **19**(3), 1985, 225–234.

[RMM+94] K. Ryall, J. Marks, M. Mazer, and S. Shieber, "Semi-automatic delineation of regions in floor plans," Harvard University, 1994.

Using the Visibility Complex for Radiosity Computation

Rachel Orti   Frédo Durand   Stéphane Rivière   Claude Puech

iMAGIS * / GRAVIR - IMAG, BP 53, F-38041 Grenoble Cedex 09, France
(Email: Rachel.Orti@imag.fr)

Abstract. The radiosity method is particularly suitable for global illumination calculations in static environments. Nonetheless, for applications of image synthesis such as lighting design or architectural simulation, we have to deal with dynamic environments. To make the method
usable in a real case, the illumination has to be updated as fast as possible after an object moves. The efficient way is to find the calculations
strictly necessary to be recomputed after a change in the scene. The
largest part of the computation time is spent on visibility calculation. In
this paper, we investigate the possible speed ups in those calculations.
We propose the use of the visibility complex for radiosity calculations.
The presented study is realized for 2D scenes of convex objects in the
static case. We show that the visibility complex is very suitable for radiosity calculations in this context, and that it also allows for efficient
updates in the dynamic case.

Keywords: radiosity, discontinuity meshing, form factor, visibility complex, dynamic environments

1   Introduction

Visibility is a very important topic in computer graphics, and especially in global
illumination calculation. Visibility computation requires an efficient method for
determining the objects seen from a point in a scene or from an object in the
scene. In the case of the radiosity method (a method for the simulation of light
interactions between the objects in a scene), the largest part of the computation
time is spent calculating the fraction of light leaving one object that arrives at
another one, since it involves many visibility calculations. To deal with the increasing complexity of the scenes considered in computer graphics, some research
has been performed in visibility processing in order to accelerate visibility determination during illumination calculations [23, 24]. The idea is to build a special
data structure that allows for easy selection of a set of potentially visible objects.

Radiosity was initially applied to static environments, currently it starts to
deal with moving objects. The work done so far still involves too many recomputations [1, 2, 5, 21], even if only a single object moves in a scene. No method has

---
* iMAGIS is a joint project of CNRS, INRIA, Institut National Polytechnique de
Grenoble and Université Joseph Fourier.

yet been proposed which exactly identifies what has to be strictly recomputed. In computational geometry, a data structure called the visibility complex has recently been developed. It deals with visibility between objects in 2D scenes. We think that this structure will allow to better understand the phenomenon of 2D dynamic environments, a step necessary for the development of efficient algorithms for radiosity in dynamic 3D scenes. Some research in lighting simulation has already been carried out in 2D for discontinuity meshing and radiosity calculations [8, 9] or wavelet radiosity approaches [20, 7]. 2D radiosity is a good laboratory for "real" 3D radiosity. The possibility of analytic computations offers a way to validate the models; the scenes are simpler to understand and allow for a better comprehension of the phenomena. Moreover, many of 3D scenes are in fact "$2\frac{1}{2}$D" scenes, like the inside of a building where the occlusions are mainly caused by the walls, which are equivalent to their projections onto a 2D plan.

In this paper, we present the use of the visibility complex for radiosity computation. We show the high suitability of this structure that helps to accurately find the mutually visible parts for all pairs of objects of the scene. Therefore, we can consider the light interactions between only mutually visible parts of objects. We can also directly compute the ratio of the light exchange between two objects, thanks to the information contained in the complex. The study is presently realized in the static case. However, we show how the visibility complex can be used to considerably improve the computation time in the dynamic case. Making the visibility complex dynamic will make possible the strictly necessary recomputations.

2 The Radiosity Method

2.1 Principles of the Method

Radiosity is a global illumination method used to render a scene of objects by computing the lighting for each point in a scene [22]. This approach is based on the physical principle of heat transfer between surfaces and simulates the balance of light energy between radiating surfaces for illumination computations.

The radiosity method was first applied to computer graphics in 1984 by Goral *et al.* [6]. With this method, the light inter-reflections between surfaces in a closed environment are modeled by a system of equations. Surfaces are assumed to be Lambertian (i.e. to reflect incident light in all directions with equal intensity). To set up the system of equations, the environment is divided into small areas (or patches). The radiosity B_i of a patch P_i is the total rate of light energy leaving this patch and is equal to the sum of emitted and reflected energies. This quantity is to be constant over a patch, and can be expressed, for an environment discretized into n patches, as:

$$B_i = E_i + \rho_i \sum_{j=1}^{n} B_j F_{ij} \,, \tag{1}$$

where E_i is the light emitted from the patch P_i, ρ_i is the reflectivity of this patch, and F_{ij} is the form factor between the patches P_i and P_j, that is the fraction of energy leaving P_i that arrives at P_j.

The accuracy of the radiosity solution depends on the discretization (or mesh) of the environment. To have good results the mesh should follow the distribution of light in the environment, with a higher density of elements in areas where the illumination changes rapidly [22]. The best method is to use a *discontinuity meshing* that places mesh boundaries on the radiosity discontinuities caused by occlusions [8] (i.e. on shadow limits). This meshing strategy traditionally requires many geometric calculations which make it very expensive to use.

Most of the computation time in the radiosity method is spent on the calculation of the form factors (still more than 50% of the time in the case of efficient algorithms, e.g. [10]) because of the visibility computations. Thus, it is important to have an efficient method to compute form factors, especially in dynamic environments where the geometry of the scene can change and causing the need for some form factors being recomputed. Below we show that the visibility complex offers a way to easily compute both the discontinuity mesh and the form factors between objects for a given scene.

2.2 The Form Factor

The form factor is the fraction of energy leaving one surface that arrives at another one [22]. The form factor F_{ij} between two surface elements A_i and A_j is defined as:

$$F_{ij} = \frac{\text{Radiative energy reaching } A_j \text{ from } A_i}{\text{Total radiative energy leaving } A_i} . \tag{2}$$

The form factor is expressed in 3D as a double integral over areas which takes into account the visibility between surfaces. In the 2D case, it becomes a double integral over lengths. The form factor between two length elements L_i and L_j corresponds to:

$$F_{ij} = \frac{1}{L_i} \int_{L_i} \int_{L_j} \frac{\cos \phi_i \cos \phi_j}{2r} H_{ij} dL_j dL_i , \tag{3}$$

where H_{ij} is the visibility between the elements dL_i and dL_j (1 if dL_i and dL_j are mutually visible, 0 otherwise) and r the distance between these elements (see Figure 1).

The form factor is, in fact, a strictly geometric quantity: it depends only on the shape and relative location of surfaces in the scene.

We will consider two other equivalent expressions for the form factor based on this property: the second being valid only in 2D.

• **A ratio of measures of lines:** as the energy is transferred through light rays, the fraction of energy corresponding to the form factor can be formulated as a ratio of measures of sets of lines. Sbert [19] has used this principle to compute an

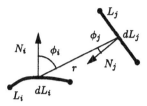

Fig. 1. Notation for the form factor.

approximation of the form factors between two surfaces in 3D. We consider here the *measure of lines* as defined within the context of Integral Geometry in [18]. Using this measure, the expression for the 2D form factor between two portions of curves C_i and C_j becomes:

$$F_{ij} = \frac{\text{Measure of lines intersecting } C_i \text{ and } C_j}{\text{Measure of lines intersecting } C_i}. \tag{4}$$

- **A weighted sum of curves lengths**: a formula called the "string rule," established in thermal engineering [11], allows for a simple computation of the 2D form factor between two portions of curves C_i and C_j by computing the length of "strings" drawn between the endpoints of C_i and C_j (see Figure 2). The strings stretched from the endpoints of C_i to the corresponding endpoints of C_j (i.e., a to c and b to d) are called *un-crossed strings*, and those drawn to the opposite endpoints on C_j (i.e., a to d and b to c) are called *crossed strings*. The form factor between two portions of curves C_i and C_j (with lengths L_i and

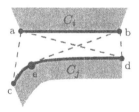

Fig. 2. Strings for two portions of curves C_i and C_j.

L_j respectively) becomes:

$$F_{ij} = \frac{\sum \mathcal{L}(crossed\ strings) - \sum \mathcal{L}(un\text{-}crossed\ strings)}{\in \mathcal{L}_)}, \tag{5}$$

where $\mathcal{L}(|)$ represents the length of the curve c.

This formulation is valid even if some parts of C_i and C_j are not mutually visible. In the case of an obstructed view, the strings are simply imagined to stretch around the obstruction (see the portion of curve between e and c). For the example given in Figure 2, the form factor F_{ij} between C_i and C_j corresponds to:

$$F_{ij} = \frac{d_e(a,d) + d_e(b,e) + d_c(e,c) - d_e(a,c) - d_e(b,d)}{2L_i} ,$$

where $d_e(m,n)$ represents the Euclidean distance between the points m and n, and $d_c(m,n)$ the Curvilinear distance between m and n.

3 The Visibility Complex

3.1 Introduction

We consider a set of n disjoint objects in the plane that represent obstacles to the propagation of either the light or a robot. The visibility graph is an important data structure in this context. For a scene of convex objects in the plane, it is made of the bitangents which do not intersect any object. But this structure is too poor to allow for global visibility queries such as maintaining a view around a point or other global visibility computation.

That is why Pocchiola and Vegter [13] introduced the *visibility complex*, a data structure for visibility of 2D scenes.

3.2 The Visibility Complex

We limit ourselves to convex objects and add an infinite "blue sky" object to the scene for the sake of coherence. We consider the set of maximal free segments of the scene, that is, segments in free space (interior of the objects removed) of maximal length. The visibility complex is a 2-dimensional cell complex which is a partition of the set of maximal free segments according to their visibility. It is composed of three types of elements (see Figure 3):

- *Vertices* are 0-d components: a vertex corresponds to a segment tangent to two objects and touching two other objects at its endpoints.
- *Edges* are 1-d components: an edge corresponds to the segments tangent to one object and touching two other objects.
- *Faces* are 2-d components: a face is the set of segments that touch the same pair of objects, i.e., that "see" the same objects.

In order to visualize these elements we consider a duality relation which associates a point with a line and vice versa. If we represent the lines in a dual space, for example $-x\sin\theta + y\cos\theta = u \mapsto (\theta, u)$, then a given object has for each θ two tangents, $(\theta, \lambda(\theta))$ and $(\theta, \mu(\theta))$. $\lambda(\theta)$ and $\mu(\theta)$ describe two curves in the dual space. Each line (θ, u) such that $\lambda(\theta) < u < \mu(\theta)$ intersects the object.

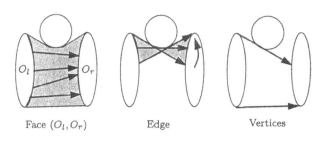

Fig. 3. Elements of the visibility complex.

For a scene of objects, these curves partition the dual space into connected components corresponding to lines intersecting the same objects. This partition is called the *dual arrangement* (see Figure 4).

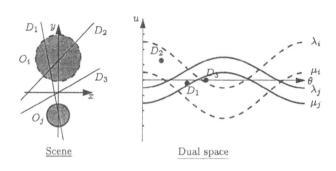

Fig. 4. Dual arrangement of a scene and some lines.

The maximal free segments can be canonically projected onto the lines, and the edges of the complex are projected onto the curves of tangency, which can help to visualize them (see Figure 5). In the dual space, we notice that:

- An *edge* is delimited by two vertices and incident to three faces.
- A *vertex* is incident to four edges and six faces.
- A *face* is delimited by two chains of vertices and edges.

The size of the complex is characterized by the number m of vertices which is $\Omega(n)$ and $\mathcal{O}(\backslash^{\epsilon})$ (with n the number of convex objects).

To have a more complete introduction to the visibility complex, refer to [13], [4] or [3].

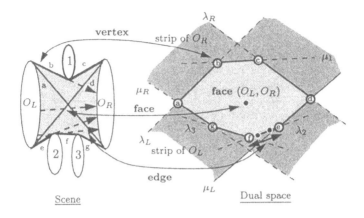

Fig. 5. Structure of a face in the scene and its projection onto the dual plane.

3.3 Construction

The construction of the complex can be achieved by a sweep of its vertices. When a vertex is swept, the relations between the edges and the faces have to be updated.

The straight sweep proposed by Pocchiola and Vegter [12], running in $\mathcal{O}(\updownarrow \log \backslash)$ time is the solution currently implemented (see [17] for implementation details).

Some care has to be taken to handle problems due to degeneracies and numerical errors (see [16]).

An implementable optimal algorithm has been proposed in [14], using a topological sweep. It runs in optimal $\mathcal{O}(\updownarrow + \backslash \log \backslash)$ time and optimal $\mathcal{O}(\backslash)$ space.

In these algorithms objects were assumed to have $\mathcal{O}(\infty)$ complexity which can be restrictive. Our current implementation computes the bitangents with a dichotomy method, which takes non negligible time. Rivire [15] has proposed and implemented an algorithm for the construction of the visibility complex of scenes of simple polygons. It runs in optimal $\mathcal{O}(\updownarrow + \backslash \log \backslash)$ time and $\mathcal{O}(\backslash)$ space, and handle the problems of degeneracies.

4 Using the Visibility Complex

4.1 Computing the Form Factors

The form factor between two elements is null if these elements are not mutually visible. In practice, such situations are frequent and should be taken into account in order to avoid unnecessary computations. Thanks to the visibility complex only pairs of mutually visible objects are considered, and only the mutually visible parts of these objects are accurately examined.

When considering the complex in the dual space (θ, u), the computation of the 2D form factor can be re-expressed by:

- **A ratio of areas of regions**:

Proposition 1. *The measure of the set of lines that intersect the object O_i is equal, in the dual space (θ, u), to the area between the two curves of tangency λ_i and μ_i for θ going from 0 to π.*

Proposition 2. *The measure of the set of lines that intersect the objects O_i and O_j is equal, in the dual space (θ, u), to the area of the corresponding face of the visibility complex.*

These propositions can be proved by using the Integral Geometry which allows to study and measure sets of lines in the plane [18]. This interpretation of measures of lines depends on the duality relation used. It is valid in duality (θ, u) because the lines are equidistributed. It is not true in duality (a, b) in which the line $l : y = ax + b$ is associated with the dual point $l^* : (a, b)$.

Using the equation 4, the form factor between two convex objects O_i and O_j can be expressed by:

$$F_{ij} = \frac{\text{Area of the face associated with } O_i \text{ and } O_j}{\text{Area between } \lambda_i \text{ and } \mu_i \text{ for } \theta \text{ going from } 0 \text{ to } \pi} \, .$$

- **A weighted sum depending on vertices and edges**:

We have seen that the form factor can be expressed by a weighted sum of curves. This expression can also be interpreted in the visibility complex. Consider the example in Figure 6. In order to determine the form factor between the objects O_i and O_j, we have to compute the lengths of curves (or "strings") stretched between the points (P_{u_l}, P_{d_r}), (P_{d_l}, P_{u_r}), (P_{u_l}, P_{u_r}), and (P_{d_l}, P_{d_r}). The straight portions of "strings" correspond to the bitangents associated with the vertices incident to the corresponding face. The curved portions of "strings" correspond to the edges bounding the face. Actually, going from P_{u_r} to P_{s_r} along the object O_j can be considered as going from D_{u_r} to D_s along the curve of tangency λ_{O_j} in the dual space.

With each vertex v of the face we associate a value $d_e(v)$ corresponding to the Euclidean length of the associated bitangent in the scene, and with each edge e of the face a value $d_c(e)$ corresponding to the curvilinear length of the corresponding portion of the associated object.

The form factor between two convex objects O_i and O_j can be expressed as:

$$F_{ij} = \frac{\sum_{v \ bounding \ face} d'_e(v) + \sum_{e \ bounding \ face} d'_c(e)}{2L_i} \, ,$$

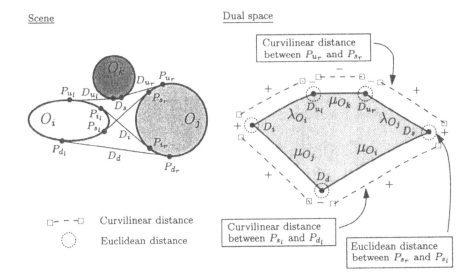

Fig. 6. Two objects O_i and O_j and their associated face in the complex.

$$\text{where } d'_e(v) = \begin{cases} d_e(v) & \text{if } v \text{ is an extremal vertex} \\ & \text{(first or last vertex)} \\ -d_e(v) & \text{otherwise} \end{cases}$$

$$\text{and } d'_c(e) = \begin{cases} d_c(e) & \text{if the object associated with } e \\ & \text{is } O_i \text{ or } O_j \\ -d_c(e) & \text{otherwise} \end{cases}$$

Figure 6 shows the different signs that modify the different values obtained for each vertex and each edge of the associated face for a given pair of objects (O_i, O_j).

Note that this method is simpler than the previous one since it does not require a numerical integration.

4.2 Computing the Discontinuity Mesh

The visibility complex allows for easy computation of the discontinuity mesh. The limits of visibility of the objects, defined by the bitangents of the scene, define discontinuities in the radiosity function.

Figure 6 shows a simple example with an obstacle that limits the visibility between the objects O_i and O_j. The bitangent D_{u_l} in the scene corresponds to the limit of the shadow caused by O_k on O_j. The intersection of this bitangent with O_j represents a point of discontinuity. A bitangent in the scene corresponds to a vertex in the complex. The set of vertices of the complex is the set of discontinuities of a scene.

Once the visibility complex is built, the discontinuity mesh can be computed, by considering the vertices of the complex, in $\mathcal{O}(\updownarrow)$ time (with m the number of vertices of the complex).

4.3 Towards an Efficient Update of the Form Factors in Dynamic Environments

In dynamic environments, two cases may occur : (1) the visibility complex is topologically unchanged: only the form factors concerning elements on the moving object must be recomputed; (2) the visibility between some objects has changed: the complex must be updated, some faces are destroyed (the corresponding form factors become null), others are created (requiring the recomputation of the corresponding form factors).

In practice, finding the form factors which need to be recomputed is not trivial. Previous applications of the radiosity method to dynamic environments, have attempted to reduce the number of unnecessary recomputed form factors ([1, 2, 5] and more recently [21]). But there is still no method exactly identifying the form factors strictly necessary to be recomputed. Making the visibility complex dynamic will make this possible.

Two types of visibility changes occur in dynamic scenes: when inserting or removing an object, and when moving an object. The first situation corresponds to the incremental construction of the visibility complex. The second modification may be handled by updating the visibility complex along the curves of tangency of the moving object. Once we have a dynamic update of the complex, we are able to identify the changes in visibility and to know exactly which form factors have to be recomputed, recomputation done efficiently with the complex. This can be illustrated with the following example.

Consider Figure 7 with an example of a moving object. When the dark gray circle is moved from the position 1 to the position 2, the face associated with the objects O_i and O_j is modified. The circle reduces the visibility between the objects O_i and O_j. A new face associated with these objects with a new position of the dark gray circle corresponds to the old face without the striped part (see Figure 7(b)). To update the form factor between the objects O_i and O_j, we just have to consider the striped part instead of recomputing everything with the new face (see Figure 8).

5 Implementation

We have implemented a system which uses the visibility complex for convex objects. It first uses the vertices of the complex for discretizing the objects of a 2D scene according to the discontinuities of visibility, and uses its faces for computing the form factors between mutually visible elements. Then the program computes the radiosity solution. It provides different visualizations: the

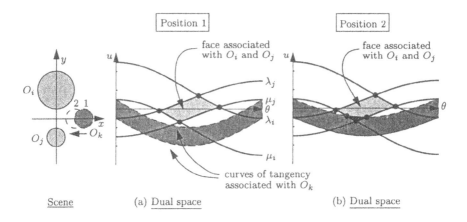

Fig. 7. Example of a circle moving from the position 1 to the position 2.

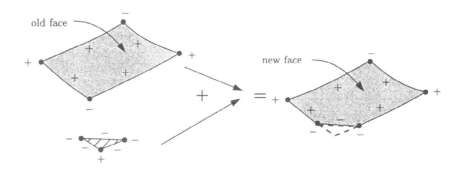

Fig. 8. Update of the form factor between two objects.

scene, the value of radiosity for each element, the matrix of form factors and the visibility complex associated to the scene.

Figure 9 shows a simple example with three circles: the circle in front is a light source which lightens the two others. Figure 9(d) illustrates that the visibility complex is not planar. The dual arrangement of the scene (see Figure 9(b)) corresponds to a projection of the complex onto the dual plane.

6 Conclusion

The study in the static case has proved that the visibility complex is very useful for radiosity computation in static scenes. The complex directly provides the

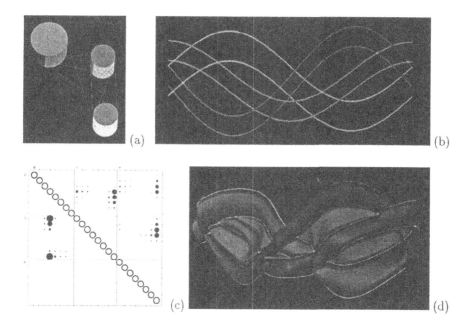

Fig. 9. (a) Scene and samples with radiosity value; (b) Dual arrangement of the scene; (c) Matrix of form factors; (d) Visibility complex of the scene.

discontinuity mesh, and prevents unnecessary computation by considering only mutually visible parts between objects. It also gives an efficient method for computing the form factors.

The visibility complex should provide important benefits for dynamic environments by offering a way of finding only the necessary re-computations. Work is currently done in this direction. We proved the visibility complex suitable for the realization of a test-bed environment for radiosity in dynamic scenes. A study in 2D makes the phenomenon of radiosity in dynamic environments more comprehensive, thus makes possible the development of efficient algorithms for radiosity in dynamic environments. We have decided to focus on the visibility complex for polygonal scenes since it is easier and more usual to work with polygonal objects in computer graphics. It is also more suitable for the hierarchical adaptations that we are studying.

For the 3D case, two approaches are being studied. For architectural scenes where the most important occlusions are vertical walls, we plan to use projections of those walls onto a horizontal plane, and to treat the visibility in the resulting 2D case. This will provide a good subset of mutually visible objects. We also envisage to build a 3D equivalent to the complex. This data-structure will have a very significant complexity, but it should have the potential to make of visibility phenomenon in dynamic environments more understandable. A hierarchical approach should decrease the complexity of such a structure.

References

1. Daniel R. Baum, John R. Wallace, Michael F. Cohen, and Donald P. Greenberg. The back-buffer algorithm: An extension of the radiosity method to dynamic environments. *The Visual Computer*, 2(5):298–306, September 1986.

2. Shenchang Eric Chen. Incremental radiosity: An extension of progressive radiosity to an interactive image synthesis system. In Forest Baskett, editor, *Computer Graphics (SIGGRAPH'90 Proceedings)*, volume 24, pages 135–144, August 1990.

3. F. Durand. Etude du complexe de visibilit. Rapport du DEA d'Informatique de Grenoble, France, June 1995.

4. F. Durand and C. Puech. The visibility complex made visibly simple. In *Proc. 11th Annu. ACM Sympos. Comput. Geom.*, page V2, 1995.

5. David W. George, Francois X. Sillion, and Donald P. Greenberg. Radiosity redistribution for dynamic environments. *IEEE Computer Graphics and Applications*, 10(4):26–34, July 1990.

6. C. Goral, K. E. Torrance, and D. P. Greenberg. Modeling the interaction of light between diffuse surfaces. In *Computer Graphics (SIGGRAPH'84 Proceedings)*, *18:3*, pages 213–222, July 1984.

7. S. J. Gortler, P. Schroder, M. F. Cohen, and P. Hanrahan. Wavelet radiosity. In *Computer Graphics (SIGGRAPH'93 Proceedings)*, pages 221–230, August 1993.

8. P. S. Heckbert. *Simulating Global Illumination Using Adaptive Meshing*. PhD thesis, UC Berkeley, June 1991.

9. P. S. Heckbert. Radiosity in flatland. In *Computer Graphics forum (EUROGRAPHICS'92 Proceedings)*, *11:3*, pages 181–192, September 1992.

10. N. Holzschuch, F. Sillion, and G. Drettakis. An efficient progressive refinement strategy for hierarchical radiosity. In *Fifth Eurographics Workshop on Rendering, Darmstadt, Germany*, pages 343–357, June 1994.

11. H. C. Hottel. Radiant heat transmission. In W. H. McAdams, editor, *Heat Transmission*, chapter 4. McGraw-Hill, New-York, 3rd edition, 1954.

12. M. Pocchiola and G. Vegter. Sweep algorithm for visibility graphs of curved obstacles. Manuscrit, Liens, Ecole Norm. Sup., Paris, June 1993.

13. M. Pocchiola and G. Vegter. The visibility complex. In *Proc. 9th Annu. ACM Sympos. Comput. Geom.*, pages 328–337, 1993.

14. M. Pocchiola and G. Vegter. Computing the visibility graph via pseudo-triangulation. In *Proc. 11th Annu. ACM Sympos. Comput. Geom.*, pages 248–257, 1995.

15. S. Rivière. Topologically sweeping the visibility complex of polygonal scenes. In *Proc. 11th Annu. ACM Sympos. Comput. Geom.*, pages C36–C37, 1995.

16. S. Rivière. Dealing with degeneracies and numerical imprecisions when computing visibility graphs. 12th European Workshop on Computational Geometry CG'96, Muenster, Germany, 1996.

17. S. Rivire. Experimental comparison of two algorithms for computing visibility graphs. Manuscrit, 1993.

18. L. A. Santalo. *Integral Geometry and Geometric Probability*, volume 1 of *Encyclopedia of Mathematics and its Applications*. Addison-Wesley Publishing Company, 1976.

19. M. Sbert. An integral geometry based method for fast form-factor computation. In *Computer Graphics forum (EUROGRAPHICS'93 Proceedings)*, *12:3*, pages 409–420, September 1993.

20. P. Schroeder, S. Gortler, M. Cohen, and Pat Hanrahan. Wavelet projections for radiosity. In *Proc. 4th Eurographics Workshop on Rendering, Paris, France*, pages 105–114, June 1993.

21. E. S. Shaw. Hierarchical radiosity for dynamic environments. Master's thesis, Cornell University, August 1994.

22. F. X. Sillion and C. Puech. *Radiosity and Global Illumination*. Morgan Kaufmann Publishers, Inc., 1994.

23. S. J. Teller. *Visibility Computations in Densely Occluded Polyhedral Environments*. PhD thesis, UC Berkeley, 1992.

24. S. J. Teller and P. M. Hanrahan. Global visibility algorithms for illumination computations. In *Computer Graphics (SIGGRAPH'93 Proceedings)*, pages 239–246, August 1993.

The CGAL Kernel: A Basis for Geometric Computation

Andreas Fabri[1], Geert-Jan Giezeman[2], Lutz Kettner[3], Stefan Schirra[4]
and Sven Schönherr[5]

[1] INRIA, B.P.93, 06902 Sophia-Antipolis cedex, France. email:
Andreas.Fabri@sophia.inria.fr.

[2] Department of Computer Science, Utrecht University, P.O. Box 80.089,3508 TB
Utrecht, The Netherlands. email: geert@cs.ruu.nl.

[3] Theoretical Computer Science, ETH Zürich, 8092 Zürich, Switzerland email:
kettner@inf.ethz.ch.*

[4] Max-Planck-Institut für Informatik, 66123 Saarbrücken, Germany. email:
stschirr@mpi-sb.mpg.de.

[5] Fachbereich Mathematik und Informatik, Freie Universität Berlin, Takustraße 9,
14195 Berlin, Germany. email: sven@inf.fu-berlin.de.**

Abstract. A large part of the CGAL-project is devoted to the development of a Computational Geometry Algorithms Library, written in C++.
We discuss design issues concerning the CGAL-kernel which is the basis
for the library and hence for all geometric computation in CGAL.

1 Introduction

Only a few of the many algorithms developed over the past two decades in
computational geometry found their way into practice. Reasons for this are the
dissimilarity between fast floating-point arithmetic normally used in practice
and exact arithmetic over the real numbers assumed in theoretical papers, the
lack of explicit handling of degenerate cases in these papers, and the inherent
complexity of many (theoretically) efficient solutions. For these reasons there is a
definite need for correct and efficient implementations of geometric algorithms.
It is the ambitious goal of the CGAL-project to provide this, i.e. to make the
large body of geometric algorithms, developed within the field of computational
geometry, available for industrial application.

In the next section we will give a short introduction to the CGAL project,
then we briefly present the kernel. In Section 4 we discuss issues related to the
design of the kernel. Section 5 gives an overview about how I/O, especially visualization, of the objects in the CGAL-kernel is handled. In Section 6 we mention
first experiences with the CGAL-kernel, Section 7 gives a short conclusion.

---

* Part of this work had been done at the graduate school "Algorithmische Diskrete
Mathematik" at FU Berlin, supported by DFG, grant We 1265/2-1.
** Supported by DFG, grant We 1265/5-1 (Leibniz-Preis).

2 The CGAL Project

CGAL is a common project of seven European sites, Utrecht University (The Netherlands), ETH Zürich (Switzerland), Free University Berlin (Germany), INRIA Sophia-Antipolis (France), MPI Saarbrücken (Germany), RISC Linz (Austria), and Tel-Aviv University (Israel). There is direct cooperation with industrial partners on concrete problems in the application domains geographical information systems, visualization and simulation, CAD/CAM, and shape analysis and reconstruction. Furthermore a library consisting of a kernel of geometric data types and operations, a large collection of geometric algorithms for 2-dimensional, 3-dimensional, and higher-dimensional problems, and support libraries for visualization, I/O, conversion to standard geometry formats used in industry, etc. will be developed within the project. Although more and more research papers in computational geometry address implementation issues of geometric algorithms (examples are [5, 6, 9, 12, 13, 15, 21]), a lot of theoretical and experimental research is still needed. Such research is also part of the CGAL-project.

3 The Kernel

The kernel has been designed in Summer 1995 by the CGAL-sites Berlin, Saarbrücken, Sophia-Antipolis and Utrecht, where the latter three already had ample experience with geometric software libraries, namely the geometric part of LEDA [17], C++GAL [2] and PLAGEO [11], respectively. The kernel is the basis of the whole library. It provides the geometric primitives underlying geometric computations. The geometric primitives in the CGAL-kernel include constructors for geometric objects such as points, planes, spheres, polyhedra, and predicates on geometric objects such as the orientation test for points, sidedness test, intersection tests and computations, and distance computations. The underlying geometry for the primitives in the CGAL-kernel is affine geometry.

The first design decision when developping a library is the choice of the programming language. We opted for C++ as as it is widely used and as it can easily be interfaced with existing C and Fortran code. It further allows to use existing libraries written in C++, for example LEDA. We consider C++ as a compromise between aesthetics and efficiency. Eiffel or Smalltalk are more properly object oriented but lack acceptance. Java is, at least for the moment, too slow due to the permanent need of runtime type identification (all functions are virtual, in C++ terminology). We will provide a bridge from Java to CGAL but it seems not interesting to us to develop the entire library in this language.

Another major design decision was to write generic code. Geometric algorithms consist of different layers, the bottommost layers being geometric primitives, i.e. basic objects and predicates, and the arithmetic used to do the calculations in the predicates. Algorithms in CGAL will be generic, i.e. work with a variety of implementations of predicates and subtasks, and representations of geometric objects. That allows to easily interchange components as long as they have the same syntax. Genericity could have been achieved by using inheritance and virtual functions or by using templates. We opted for writing template code as it has the advantage that it incurs no extra runtime costs.

An example for genericity of the CGAL-kernel is the arithmetic. Templates are used to allow the user to choose the underlying arithmetic. One can choose between representations of the geometric objects based on representation of points by Cartesian (or affine) coordinates and representations based on homogeneous coordinates. The latter allows to reduce many computations involved in geometric algorithms to calculations over the integers, since divisions can be avoided with this representation. The representation based on Cartesian coordinates is called C-representation, the homogeneous counterpart is called H-representation. The C-representation is parameterized by a number type NT which is used in all computations in the geometric primitives involving objects parameterized with representation C<NT>. The H-representation is parameterized by two number types FT (field type) and RT (ring type). In the representation H<FT,RT> the number type RT is used for all internal computations and is typically a number type representing integers, whereas FT is used whenever the computations involve operations beyond $+, -, *$. For example, FT is used when you ask for the Cartesian x-coordinate of a point, since here division by the homogenizing coordinate is needed. Note that the H-representation is also for affine geometry.

The set of geometric objects in the kernel is closed under affine transformations. For example, the geometric object sphere induces that the class ellipse is in the CGAL-kernel as well. Points, vectors, and plane equations behave differently under affine transformations [20]. We use the strong type system and operator overloading of C++ to hide these details of different transformation properties in the transformation class. Additionally, the type checking detects the incorrect use of operations like the addition of two points. No type cast is provided to convert between points p and vectors v. Instead, a symbolic origin o is introduced and the expressions $v = p - o$ and $p = o + v$ are legal. Internally, operator overloading maps these calls to hidden type casts, so efficiency is guaranteed.

The C- and H-representation provide a powerful mechanism to support multiple representations. In fact, they are just examples of two different representations for coordinate values. In principle, there exists a general *interface* class for each object, e.g. CGAL_Plane_3 for a three dimensional plane, that is a template parameterized with the C- or H-representation. The representation class itself contains a mapping from the interface class to the specific *implementation* class. The mapping is implemented via a typedef. In the following example for the C-representation the mapping goes to the specific CGAL_PlaneC3 class which implements a plane.

```
template < class NT>
class C {
  typedef CGAL_PlaneC3<NT>   Plane_3;
  ...
}
```

Note that both CGAL_PlaneC3 and the C-representation are parameterized by the number type NT. The interface class is now implemented as follows:

```
template < class R>
class CGAL_Plane_3 : public R::Plane_3 {
    ...
}
```

The example in Fig. 1 shows how a search in a binary space partition tree [7, pp. 555-557] could be implemented using CGAL. It serves as an example of what CGAL-code looks like, not of how a BSP tree should be implemented. Of course, using the higher levels of CGAL there should be no need to implement a BSP tree.

```
template <class R>
struct bsp_cell {
    CGAL_Line_2<R> splitter;
    bsp_cell<R> *left;
    bsp_cell<R> *right;
    bool is_leaf;
}

template <class R>
bsp_cell<R> *
search_cell(bsp_cell<R> *cur,
            CGAL_Point_2<R> point)
{
    if (cur→is_leaf)
        return cur;
    switch(cur→splitter.where_is(point) {
      case CGAL_LEFT:
        return search_cell(cur→left, point);
      case CGAL_RIGHT:
        return search_cell(cur→right, point);
      case CGAL_ON:
        return cur;
    }
}
```

Fig. 1. An example of CGAL-code

4 Design Goals

This section describes goals we had in mind when we designed the CGAL-kernel. Robustness, efficiency, generality, and ease of use are the primary design goals.

4.1 Robustness

What is robustness? Dey et al. [5] define robustness as the ability of a geometric algorithm to deal with degeneracies and the inaccuracies during various numerical computations. Fortune [8] calls an algorithm robust if it always computes an answer that is the correct answer for some perturbation of the input. He calls it stable if that perturbation is small. Sometimes an algorithm is already called robust if it does not fail. A safe way to robustness is exact computation. It simply makes robustness a non-issue and preserves all geometric concepts [21].

Exact geometric primitives ensure exact geometric computation. For many problems (cf. [21]) the C-representation together with exact rational arithmetic, e.g. the number type **rational** from LEDA [17], or the H-representation together with exact integer arithmetic, e.g. BigNum [18] or LEDA's number type **integer**, as RT number type lead to exact geometric predicates. For exact geometric computation it is, however, not necessary that all computations are done exactly. Geometric predicates often just encapsulate decisions based on numerical computations. It is often sufficient to determine the sign of some expression E, the exact value of E is not needed. Lazy evaluation [4] and floating-point filters [9] can be used to do exact geometric computation much more efficiently. If these techniques are incorporated in a number type only, the computation history has to be managed, e.g., by storing expression dags. Other useful number types are **real** in LEDA and those described in [21].

Parameterization by number types allows to create special versions for geometric predicates by defining a new number type and specializing[3] the predicates for this number type. This allows to use floating-point filters on the level of geometric predicates. Having information on the size of the numerical data involved in a geometric predicate, a specialization of the predicate can be coded that uses static floating-point filters. Tools like Fortune and van Wyk's LN [10] can be used to generate code for such a specialization. Many geometric predicates are pure sign computations of determinants. Recent results on the sign computation of small determinants of a matrix with integral **double**-entries [3] are incorporated in the CGAL-kernel by specializing appropriate predicates for a number type that encapsulates integral doubles.

4.2 Efficiency

Efficiency is an absolute must. A library providing inefficient algorithms will not be used. In some applications efficiency seems to be even more important than exactness. Therefore the CGAL-kernel will offer performance-optimized robust specializations of geometric primitives for fast floating-point number types.

Here exactness is no longer guaranteed. However, the CGAL-kernel will guarantee correctness, in the sense that the predicates come with a precise description

---

[3] Specializing means writing code for functions and classes that would otherwise be automatically generated by the template instantiation mechanism. In C++ terminology the former is called explicit specialization, the latter implicit specialization.

for which inputs they work and what they compute. This is not true for many implementations of geometric algorithms. Consequently users are often disappointed by unexpected behavior of the (implementation of an) algorithm, in the extreme case by unexpected failures and break downs. There are many faces of correctness. An algorithm handling only non-degenerate cases can be correct in the above sense. Also, an algorithm that guarantees to compute the exact result only if the numerical input data are integral and smaller than some given bound can be correct as well as an algorithm that computes an approximation to the exact result with a guaranteed error bound. One of the goals of the CGAL-kernel will be to combine efficiency and exact geometric computation.

Here is an example illustrating how efficiency is affected by the kernel design. The computation of the intersection of two objects happens in two steps: first, the type of the resulting object is computed, then the intersection object is computed. This two step computation is necessary due to the strong typing of C++, but as the second step would redo most of the computations done in the first step we store an intermediate result. Note that this allows arbitrarily complex intermediate results, e.g., when two polygons are intersected it might be a list of segments and a list of points.

Another example are affine transformations. The general form of affine transformations is too time- and space-inefficient to model restricted transformations such as translation, rotation, or scaling. The CGAL-kernel will provide specialized representations for these transformations, all hidden behind a unified interface following the design goal *ease of use*.

4.3 Generality

The applications of the CGAL-library will be very heterogeneous and hence also the applications of the geometric primitives in the CGAL-kernel. Clearly a library algorithm cannot be the best solution for every application but it can be close to it. In this sense it can be general. The parameterization by representation and the resulting possibility to specialize offer a lot of choices, e.g. by choosing an appropriate number type (and specializing geometric predicates) you can trim CGAL-algorithms for a concrete application.

4.4 Ease of Use

The kernel and later the basic library are a unified whole. The naming scheme is uniform, for the classes describing geometric objects as well as for the member functions of these classes.

In a geometric software system the kernel of a library is the bottom-most layer. The correctness of the higher layers clearly relies on the correctness of the routines provided by the kernel. The routines in the CGAL-kernel are designed for checkability and (by default) self-checking. For example, pre-conditions of routines are checked and results of computations are "verified" by checking post-conditions. These checkings are of great help in the implementation process and can reduce debugging efforts drastically. They are also an important prerequisite for program checking at higher levels [14]. Checking can be switched off of course, e.g. when code goes in production mode.

4.5 Modularity

Usually a geometric algorithm can be decomposed into subtasks until finally the level of geometric primitives, provided by the CGAL-kernel, is reached. For many of these subtasks CGAL-algorithms will become available. Using these algorithms and geometric primitives from the CGAL-kernel the parts will work together in a seamless way.

Robustness is directly related to modularity. Most of the time it is easy to ensure robustness for a single geometric predicate: since the computation depth of the predicates is small, backwards analysis of the numerical computation gives bounds on the perturbation of the numerical data of the objects involved in the predicate. Combination of predicates however, makes life complicated. Different geometric predicates might require different perturbations and the consistent combination of the outcomes of the predicates to a consistent geometric computation might be impossible. This is one of the reasons why geometric primitives in the kernel aim for perfect robustness, i.e. exactness.

4.6 Openness

The fact that the entire library is templated by numbertypes, allows you to add your own geometric predicates to the CGAL-kernel. Via specializations through new number types already existing predicates in the library can be substituted by alternative implementations. A typical example is determinant computation: if you can exploit properties of a certain numbertype to develop a faster or more exact determinant function, you can tell the compiler not to specialize the template but to take your code instead.

For many geometric objects there are different ways to implement them. In Section 3 we have shown how interface classes are mapped to implementation classes. This mechanism even allows to plug in alternative implementation classes. For example an alternative plane implementation class My_PlaneC3 can be added by deriving a new representation class from C<NT> and by overriding the typedef CGAL_PlaneC3 Plane_3.

```
template < class NT>
class My_C : public C<NT> {
   typedef My_PlaneC3<NT>   Plane_3;
}
```

The benefits are: the scheme is simple to use, user extendible, and multiple representations can coexist in an application. A crucial point is the assignment for objects that have the same representation in different representation classes. In the example above, the representation of a point is not affected by the change in the representation of a plane. Both CGAL_Point_3<C<NT> > and CGAL_Point_3<My_C<NT> > are actually derived from CGAL_PointC3<NT>. A down cast from the base class into the interface class CGAL_Point_3<R> makes a proper assignment between them possible.

We have just shown how the user can change implementations of a geometric class without affecting code that makes use of this class. However, this approach does not easily allow to switch between different implementations. For the example we have given, an assignment between the two plane types is only possible if the operator My_PlaneC3<NT>::operator=(CGAL_PlaneC3<NT>&) is defined.

Different implementations mainly means different internal representations: a line might be represented by storing the coefficients of its equation or by storing two points. The purpose of using classes is just to hide the representation as an implementation detail, but under certain circumstances it is desirable that a user can choose the internal representation: converting to a canonical representation can introduce imprecision if a numbertype as double is used.

Currently we investigate derivation from an abstract base class to solve the multiple representation problem. Different representations are put in different derived classes, see Figure 2. All member functions for which the implementation depends on the internal representation are virtual. All this is hidden to the user, who only sees *handles*, that is objects containing a pointer to representation objects. The base class Line only provides default constructor, copy constructor, and assignment operator. It further passes all member function calls through to the representation object it points at. The C++ run time system calls the appropriate virtual function. Derived classes as Line_eq provide constructors that create the appropriate representation object.

Note that Line is not an abstract base class. An object of class Line can be a placeholder for any object from a derived line class. This allows to write generic code which is best illustrated by an example:

```
void fct(const Line &l)
{
  Line h = l.opposite();
  Line_eq leq = l.opposite();
  ...
}
```

The above function can be called with an object from any class derived from class Line. Variable h will hold an object of the same type as 1. Variable leq will hold a line in equation form, regardless of the type of 1. If necessary a conversion takes place.

This approach to multiple representations is open in the sense that a user can simply add further representations. The main disadvantages are that virtual function calls take more time than ordinary function calls and that the handle objects introduce a further level of indirection, when we access data in the representation objects.

Another point where the library is open, is how it deals with collections of objects. We adopt the iterator approach of the C++ Standard Template Library [19, 16], a part of the forthcoming C++ standard [1]. An iterator can be as simple as a pointer into an array, but it can also iterate over the elements of a linked

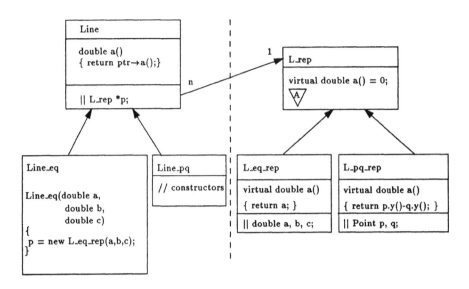

Fig. 2. Multiple representations. The user sees the class hierarchy on the left hand side only.

list or traverse a tree. This makes that the user can store data, for example the points of a polygon, in an arbitrary container. There is no need to copy them to a default container before they can be used by CGAL objects.

5 Visualization

Algorithms in CGAL, and hence objects in the CGAL-kernel need interfaces to the outside world. Users want to display their results with various kinds of *visualization* tools. During the development of CGAL people need tools which allow graphical debugging of their algorithms. Graphical debugging here means, that in a *dbx* session you can obtain a pointer on the internal representation of an object by clicking on its graphical representation in the visualization tool. This pointer can be used as an entry point in a linked data structure, e.g. a planar graph describing the convex hull of points in 3D space, which can then be explored in the debugger.

To keep I/O out of the kernel was a design decision, motivated by the many visualization tools and file formats that exist. It would not really make sense to favor one of them, although we will provide some implementations of I/O.

In providing this functionality we aim for *uniformity* and *openness*. The first means that from a CGAL point of view all visualizers look the same. We achieve this goal by expressing all kinds of I/O in terms of C++ streams. Openness means that a user with still another user interface or a different format for representing geometric data can easily add it to the support library. The conversion routines must be encapsulated in a new stream class for this visualizing/formatting tool. This concept can also be used to interface the library to existing software.

In the following we give two examples of how this looks like, first for output to a Postscript file or a previewer, then I/O with *geomview*.

5.1 Postscript

The ideas sketched above are best illustrated by a piece of code:

```
{
  typedef CGAL_Point_2< C<double> > Point;
  typedef CGAL_Line_2< C<double> >  Line;

  Point p(2.0, 1.0), q(10.0, 20.0);
  Line line(p,q);
  CGAL_Bbox_2 bbox(0.0, 0.0, 3.0, 3.0);
  Postscript_stream stream(bbox, "file.ps");

  stream ≪ set_line_width(2) ≪ line ≪ RED ≪ p ≪ GREEN ≪ q;
}
```

The constructor of this stream class opens a file named **file.ps** and writes the postscript preamble in that file. The constructor further has a bounding box as argument against that objects are clipped. Inserting a color in the stream means that the geometric objects that follow will be drawn in that color. Manipulators, as set_line_width, allow to modify the state of the stream. Inserters for geometric objects write Postscript expressions to the file. As soon as the stream object goes out of scope the destructor closes the file. If no file is specified, the output goes on a Postscript previewer.

5.2 Geomview

This 3D visualization tool was developed at the Geometry Center, Minneapolis. We provide bidirectional communication between an application and *geomview*. This allows to display geometric objects, to interactively add input data in *geomview* which are passed to the application, and to do graphical debugging.

```
{
  typedef CGAL_Point_3< C<double> > Point;
  typedef CGAL_Line_3< C<double> >  Line;
  Point p(2.0, 1.0, 0.0), q(10.0, 20.0, 2.0);
  Line line(p,q), *lptr;
  CGAL_Bbox_3 bbox(0.0, 0.0, 0.0, 3.0, 3.0, 3.0);
  Geomview_stream stream(bbox, "silicon.inria.fr");
```

```
stream ≪ Color(0.1, 0.6, 1.0, 0.9) ≪ line ≪ RED ≪ p ≪ GREEN ≪ q;

stream ≫ lptr; /* click on line => *lptr == line */
}
```

The constructor gets a bounding box and a name of a machine as an argument. The first argument allows to position the camera, the second argument to run *geomview* on a remote machine. It happens in the constructor that geomview is started and that UNIX pipes for the interprocess communication are set up. The idea of running *geomview* on a remote machine allows to exploit dedicated hardware of fast graphics workstations. The output operators for geometric objects construct expressions in the *geomview command language* which are sent to geomview via a pipe. The constructor of the color class has a fourth argument defining the transparency.

The geomview stream also comes with input operators for geometric objects. More important, we provide input operators for pointers on geometric objects which allows graphical debugging. The last statement in the above piece of code allows to click on an object that was displayed previously. Because the object `stream` stores pointers to all objects that where inserted in the stream it is an easy exercise to return the pointer on that object. Finally, when the stream object goes out of scope, its destructor stops *geomview*.

6 Experiences

We still have very little experience with the CGAL-kernel. It is currently used by students at FU Berlin, Saarbrücken University, Utrecht University and INRIA Sophia-Antipolis. Their reactions will certainly lead to modifications of the CGAL-kernel.

7 Conclusion

We hope that due to its versatility, efficiency, and robustness the CGAL-kernel will be a basis not only for geometric computation in the CGAL-project, but also for many projects involving implementation of geometric algorithms, from rapid prototyping of geometric software to final software products in the application areas of computational geometry.

A software library lives by its acceptance. Hence it is important to present the design of its basis and get reactions at an early stage when it is still fairly easy to correct it.

References

1. ANSI/ISO C++ Standards Committee *Working Paper for Draft Proposed International Standard for Information Systems – Programming Language C++*. Doc

202

No: X3J16/95-0087, WG21/N0687. April 1995.
ftp://research.att.com:/dist/c++std/WP

2. F. Avnaim. C++GAL: A C++ library for geometric algorithms. 1994.

3. F. Avnaim, J.D. Boissonnat, O. Devillers, F.P. Preparata, and M. Yvinec. Evaluating signs of determinants using single precision arithmetic. Technical Report 2306, INRIA Sophia-Antipolis, 1994.

4. M.O. Benouamer, P. Jaillon, D. Michelucci, and J-M. Moreau. A "lazy" solution to imprecision in computational geometry. In *Proc. of the 5th Canadian Conference on Computational Geometry*, pages 73–78, 1993.

5. T.K. Dey, K. Sugihara, and C.L. Bajaj. Delaunay traingulations in three dimensions with finite precision arithmetic. *Computer Aided Geometric Design*, 9:457–470, 1992.

6. D. Dobkin and D. Silver. Applied computational geometry: Towards robust solutions of basic problems. *Journal of Computer and System Sciences*, 40:70–87, 1990.

7. J. D. Foley, A. van Dam, S. K. Feiner, and J. F. Hughes. *Computer Graphics: Principles and Practice*. Addison-Wesley, Reading, MA, 1990.

8. S. Fortune. Stable maintenance of point-set triangulations in two dimensions. In *Proceedings of the 30th IEEE Symposium on Foundations of Computer Sience*, pages 494–499, 1989.

9. S. Fortune and C. van Wyk. Efficient exact arithmetic for computational geometry. In *Proc. of the 9th ACM Symp. on Computational Geometry*, pages 163–172, 1993.

10. S. Fortune and C. van Wyk. LN user manual. 1993.

11. G.-J. Giezeman. PlaGeo, a library for planar geometry, and SpaGeo, a library for spatial geometry. 1994.

12. L. Guibas, D. Salesin, and J. Stolfi. Epsilon geometry: Building robust algorithms from imprecise computations. In *Proc. of the 5th ACM Symp. on Computational Geometry*, pages 208–217, 1989.

13. K. Mehlhorn and S. Näher. The implementation of geometric algorithms. In *13th World Computer Congress IFIP94*, volume 1, pages 223–231. Elsevier Science B.V. North-Holland, Amsterdam, 1994.

14. K. Mehlhorn, S. Näher, T. Schilz, S. Schirra, R. Seidel, M. Seel, and C. Uhrig. Checking geometric programs or verification of geometric structures. Proceedings of *12th Annual ACM Symp. on Computational Geometry*, pages 159–165, 1996.

15. V. Milenkovic. Verifiable implementations of geometric algorithms using finite precision arithmetic. *Artificial Intelligence*, 37:377–401, 1988.

16. D.R. Musser, A. Saini. *STL Tutorial and Reference Guide*. Addison-Wesley, 1996.

17. S. Näher and C. Uhrig. The LEDA User Manual, Version R 3.2. Technical Report MPI-I-95-1-002, Max-Planck-Institut für Informatik, 1995.

18. B. Serpette, J. Vuillemin, and J.C. Hervé. BigNum, a portable and efficient package for arbitrary-precision arithmetic. Technical Report Research Report 2, Digital Paris Research Laboratory, 1989.

19. A. Stepanov, M. Lee. *The Standard Template Library*. July 1995.
http://www.cs.rpi.edu/~musser/stl.html

20. Ken Turkowski. Properties of surface-normal transformations. In Andrew S. Glassner, editor, *Graphics Gems*, pages 539–547. 1990.

21. C.K. Yap. Towards exact geometric computation. In *Proc. of the 5th Canadian Conference on Computational Geometry*, pages 405–419, 1993.

Triangle: Engineering a 2D Quality Mesh Generator and Delaunay Triangulator

Jonathan Richard Shewchuk

School of Computer Science, Carnegie Mellon University,
Pittsburgh, Pennsylvania 15213, USA

Abstract. Triangle is a robust implementation of two-dimensional constrained Delaunay triangulation and Ruppert's Delaunay refinement algorithm for quality mesh generation. Several implementation issues are discussed, including the choice of triangulation algorithms and data structures, the effect of several variants of the Delaunay refinement algorithm on mesh quality, and the use of adaptive exact arithmetic to ensure robustness with minimal sacrifice of speed. The problem of triangulating a planar straight line graph (PSLG) without introducing new small angles is shown to be impossible for some PSLGs, contradicting the claim that a variant of the Delaunay refinement algorithm solves this problem.

1 Introduction

Triangle is a C program for two-dimensional mesh generation and construction of Delaunay triangulations, constrained Delaunay triangulations, and Voronoï diagrams. Triangle is fast, memory-efficient, and robust; it computes Delaunay triangulations and constrained Delaunay triangulations exactly. Guaranteed-quality meshes (having no small angles) are generated using Ruppert's Delaunay refinement algorithm. Features include user-specified constraints on angles and triangle areas, user-specified holes and concavities, and the economical use of exact arithmetic to improve robustness. Triangle is freely available on the Web at "http://www.cs.cmu.edu/~quake/triangle.html" and from Netlib. This paper discusses many of the key implementation decisions, including the choice of triangulation algorithms and data structures, the steps taken to create and refine a mesh, a number of issues that arise in Ruppert's algorithm, and the use of exact arithmetic.

2 Triangulation Algorithms and Data Structures

A triangular mesh generator rests on the efficiency of its triangulation algorithms and data structures, so I discuss these first. I assume the reader is familiar with Delaunay triangulations, constrained Delaunay triangulations, and the incremental insertion algorithms for constructing them. Consult the survey by Bern and Eppstein [2] for an introduction.

There are many Delaunay triangulation algorithms, some of which are surveyed and evaluated by Fortune [7] and Su and Drysdale [18]. Their results indicate a rough parity in speed among the incremental insertion algorithm of Lawson [11], the divide-and-conquer algorithm of Lee and Schachter [12], and the plane-sweep algorithm of Fortune [6]; however, the implementations they study were written by different people. I believe that Triangle is the first instance in which all three algorithms have been implemented with the same data structures and floating-point tests, by one person who gave roughly equal attention to optimizing each. (Some details of how these implementations were optimized appear in Appendix A.)

Table 1 compares the algorithms, including versions that use exact arithmetic (see Sect. 4) to achieve robustness, and versions that use approximate arithmetic and are hence faster but may fail or produce incorrect output. (The robust and non-robust versions are otherwise identical.) As Su and Drysdale [18] also found, the divide-and-conquer algorithm is fastest, with the sweepline algorithm second. The incremental algorithm performs poorly, spending most of its time in point location. (Su and Drysdale produced a better incremental insertion implementation by using bucketing to perform point location, but it still ranks third. Triangle does not use bucketing because it is easily defeated, as discussed in the appendix.) The agreement between my results and those of Su and Drysdale lends support to their ranking of algorithms.

An important optimization to the divide-and-conquer algorithm, adapted from Dwyer [5], is to partition the vertices with alternating horizontal and vertical cuts (Lee and Schachter's algorithm uses only vertical cuts). Alternating cuts speed the algorithm and, when exact arithmetic is disabled, reduce its likelihood of failure. One million points can be triangulated correctly in a minute on a fast workstation.

All three triangulation algorithms are implemented so as to eliminate duplicate input points; if not eliminated, duplicates can cause catastrophic failures. The sweepline algorithm can easily detect duplicate points as they are removed from the event queue (by comparing each with the previous point removed from the queue), and the incremental insertion algorithm can detect a duplicate point after point location. The divide-and-conquer algorithm begins by sorting the points by their x-coordinates, after which duplicates can be detected and removed. This sorting step is a necessary part of the divide-and-conquer algorithm with vertical cuts, but not of the variant with alternating cuts (which must perform a sequence of median-finding operations, alternately by x and y-coordinates). Hence, the timings in Table 1 for divide-and-conquer with alternating cuts could be improved slightly if one could guarantee that no duplicate input points would occur; the initial sorting step would be unnecessary.

Should one choose a data structure that uses a record to represent each edge, or one that uses a record to represent each triangle? Triangle was originally written using Guibas and Stolfi's *quad-edge* data structure [10] (without the *Flip* operator), then rewritten using a triangle-based data structure. The quad-edge data structure is popular because it is elegant, because it simultaneously represents a graph and its geometric dual (such as a Delaunay triangulation and the corresponding Voronoï diagram), and because Guibas and Stolfi give detailed pseudocode for implementing the divide-and-conquer and incremental Delaunay algorithms using quad-edges.

Despite the fundamental differences between the data structures, the quad-edge-based and triangle-based implementations of Triangle are both faithful to the Delaunay triangulation algorithms presented by Guibas and Stolfi [10] (I did not implement a quad-edge sweepline algorithm), and hence offer a fair comparison of the data structures. Perhaps the most useful observation of this

Table 1. Timings for triangulation on a DEC 3000/700 with a 225 MHz Alpha processor, not including I/O. Robust and non-robust versions of the Delaunay algorithms triangulated points chosen from one of three different distributions: uniformly distributed random points in a square, random approximately cocircular points, and a tilted 1000 × 1000 square grid.

Delaunay triangulation timings (seconds)						
Number of points	10,000			100,000		
Point distribution	Uniform	Boundary	Tilted	Uniform	Boundary	Tilted
Algorithm	Random	of Circle	Grid	Random	of Circle	Grid
Div&Conq, alternating cuts						
robust	0.33	0.57	0.72	4.5	5.3	5.5
non-robust	0.30	0.27	0.27	4.0	4.0	3.5
Div&Conq, vertical cuts						
robust	0.47	1.06	0.96	6.2	9.0	7.6
non-robust	0.36	0.17	failed	5.0	2.1	4.2
Sweepline						
non-robust	0.78	0.62	0.71	10.8	8.6	10.5
Incremental						
robust	1.15	3.88	2.79	24.0	112.7	101.3
non-robust	0.99	2.74	failed	21.3	94.3	failed

Number of points	1,000,000		
Point distribution	Uniform	Boundary	Tilted
Algorithm	Random	of Circle	Grid
Div&Conq, alternating cuts			
robust	58	61	58
non-robust	53	56	44
Div&Conq, vertical cuts			
robust	79	98	85
non-robust	64	26	failed
Sweepline			
non-robust	147	119	139
Incremental			
robust	545	1523	2138
non-robust	486	1327	failed

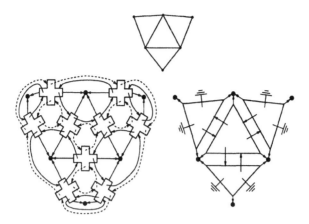

Fig. 1. A triangulation (top) and its corresponding representations with quad-edge and triangular data structures. Each quad-edge and each triangle contains six pointers.

paper for practitioners is that the divide-and-conquer algorithm, the incremental algorithm, and the Delaunay refinement algorithm for mesh generation were all sped by a factor of two by the triangular data structure. (However, it is worth noting that the code devoted specifically to triangulation is roughly twice as long for the triangular data structure.) A difference so pronounced demands explanation.

First, consider the different storage demands of each data structure, illustrated in Fig. 1. Each quad-edge record contains four pointers to neighboring quad-edges, and two pointers to vertices (the endpoints of the edge). Each triangle record contains three pointers to neighboring triangles, and three pointers to vertices. Hence, both structures contain six pointers.[1] A triangulation contains roughly three edges for every two triangles. Hence, the triangular data structure is more space-efficient.

It is difficult to ascertain with certainty why the triangular data structure is superior in time as well as space, but one can make educated inferences. When a program makes structural changes to a triangulation, the amount of time used depends in part on the number of pointers that have to be read and written. This amount is smaller for the triangular data structure; more of the connectivity information is implicit in each triangle. Caching is improved by the fact that fewer structures are accessed. (For large triangulations, any two adjoining quad-edges or triangles are unlikely to lie in the same cache line.)

---

[1] Both the quad-edge and triangle data structures must store not only pointers to their neighbors, but also the *orientations* of their neighbors, to make clear how they are connected. For instance, each pointer from a triangle to a neighboring triangle has an associated orientation (a number between zero and two) that indicates which edge of the neighboring triangle is contacted. An important space optimization is to store the orientation of each quad-edge or triangle in the bottom two bits of the corresponding pointer. Thus, each record must be aligned on a four-byte boundary.

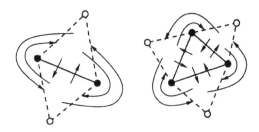

Fig. 2. How the triangle-based divide-and-conquer algorithm represents an isolated edge (left) and an isolated triangle (right). Dashed lines represent ghost triangles. White vertices all represent the same "vertex at infinity"; only black vertices have coordinates.

Because the triangle-based divide-and-conquer algorithm proved to be fastest, it is worth exploring in some depth. At first glance, the algorithm and data structure seem incompatible. The divide-and-conquer algorithm recursively halves the input vertices until they are partitioned into subsets of two or three vertices each. Each subset is easily triangulated (yielding an edge, two collinear edges, or a triangle), and the triangulations are merged together to form larger ones. If one uses a degenerate triangle to represent an isolated edge, the resulting code is clumsy because of the need to handle special cases. One might partition the input into subsets of three to five vertices, but this does not help if the points in a subset are collinear.

To preserve the elegance of Guibas and Stolfi's presentation of the divide-and-conquer algorithm, each triangulation is surrounded with a layer of "ghost" triangles, one triangle per convex hull edge. The ghost triangles are connected to each other in a ring about a "vertex at infinity" (really just a null pointer). A single edge is represented by two ghost triangles, as illustrated in Fig. 2.

Ghost triangles are useful for efficiently traversing the convex hull edges during the merge step. Some are transformed into real triangles during this step; two triangulations are sewn together by fitting their ghost triangles together like the teeth of two gears. (Some edge flips are also needed. See Fig. 3.) Each merge step creates only two new triangles; one at the bottom and one at the top of the seam. After all the merge steps are done, the ghost triangles are removed and the triangulation is passed on to the next stage of meshing.

Precisely the same data structure, ghost triangles and all, is used in the sweepline implementation to represent the growing triangulation (which often includes dangling edges). Details are omitted.

Augmentations to the data structure are necessary to support the constrained triangulations needed for mesh generation. Constrained edges are edges that may not be removed in the process of improving the quality of a mesh, and hence may not be flipped during incremental insertion of a vertex. One or more constrained edges collectively represent an input segment. Constrained edges may carry additional information, such as boundary conditions for finite element simulations. (A future version of Triangle may support curved segments this way.)

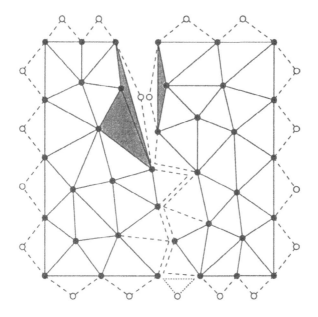

Fig. 3. Halfway through a merge step of the divide-and-conquer algorithm. Dashed lines represent ghost triangles and triangles displaced by edge flips. The dotted triangle at bottom center is a newly created ghost triangle. Shaded triangles are non-Delaunay and will be displaced by edge flips.

The quad-edge structure supports such constraints easily; each quad-edge is simply annotated to mark the fact that it is constrained, and perhaps annotated with extra information. It is more expensive to represent constraints with the triangular structure; I augment each triangle with three extra pointers (one for each edge), which are usually null but may point to *shell edges*, which represent constrained edges and carry additional information. This eliminates the space advantage of the triangular data structure, but not its time advantage. Triangle uses the longer record only if constraints are needed.

3 Ruppert's Delaunay Refinement Algorithm

Ruppert's algorithm for two-dimensional quality mesh generation [15] is perhaps the first theoretically guaranteed meshing algorithm to be truly satisfactory in practice. It produces meshes with no small angles, using relatively few triangles (though the density of triangles can be increased under user control) and allowing the density of triangles to vary quickly over short distances, as illustrated in Fig. 4. (Chew [3] independently developed a similar algorithm.) This section describes Ruppert's Delaunay refinement algorithm as it is implemented in Triangle.

Triangle's input is a *planar straight line graph* (PSLG), defined to be a collec-

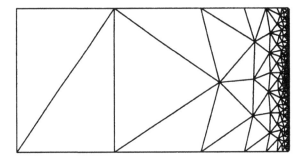

Fig. 4. A demonstration of the ability of the Delaunay refinement algorithm to achieve large gradations in triangle size while constraining angles. No angles are smaller than 24°.

tion of vertices and segments (where the endpoints of every segment are included in the list of vertices). Figure 5 illustrates a PSLG defining an electric guitar. Although the definition of "PSLG" normally disallows segment intersections (except at segment endpoints), Triangle can detect segment intersections and insert vertices.

The first stage of the algorithm is to find the Delaunay triangulation of the input vertices, as in Fig. 6. In general, some of the input segments are missing from the triangulation; the second stage is to insert them. Triangle can force the mesh to conform to the segments in one of two ways, selectable by the user. The first is to insert a new vertex corresponding to the midpoint of any segment that does not appear in the mesh, and use Lawson's incremental insertion algorithm to maintain the Delaunay property. The effect is to split the segment in half, and the two resulting subsegments may appear in the mesh. If not, the procedure is repeated recursively until the original segment is represented by a linear sequence of constrained edges in the mesh.

The second choice is to simply use a constrained Delaunay triangulation (Fig. 7). Each segment is inserted by deleting the triangles it overlaps, and retriangulating the regions on each side of the segment. No new vertices are inserted. For reasons explained in Sect. 3.1, Triangle uses the constrained Delaunay triangulation by default.

The third stage of the algorithm, which diverges from Ruppert [15], is to remove triangles from concavities and holes (Fig. 8). A hole is simply a user-specified point in the plane where a "triangle-eating virus" is planted and spread by depth-first search until its advance is halted by segments. (This simple mechanism saves both the user and the implementation from a common outlook wherein one must define oriented curves whose insides are clearly distinguishable from their outsides. Triangle's method makes it easier to treat holes and internal boundaries in a unified manner.[2]) Concavities are recognized from un-

---

[2] I imagine computational geometers replying, "Of course," engineers responding, "Hmm," and solid modeling specialists recoiling in horror.

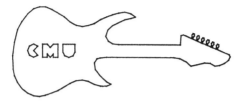

Fig. 5. Electric guitar PSLG.

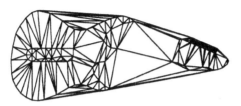

Fig. 6. Delaunay triangulation of vertices of PSLG. The triangulation does not conform to all of the input segments.

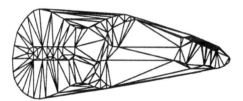

Fig. 7. Constrained Delaunay triangulation of PSLG.

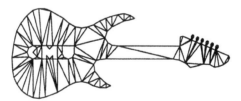

Fig. 8. Triangles are removed from concavities and holes.

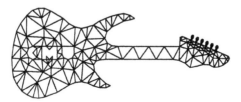

Fig. 9. Conforming Delaunay triangulation with 20° minimum angle.

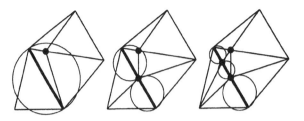

Fig. 10. Segments are split recursively (while maintaining the Delaunay property) until no segments are encroached.

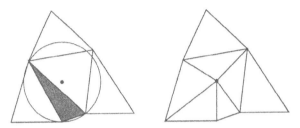

Fig. 11. Each bad triangle is split by inserting a vertex at its circumcenter and maintaining the Delaunay property.

constrained edges on the boundary of the mesh, and the same virus is used to hollow them out.

The fourth stage, and the heart of the algorithm, refines the mesh by inserting additional vertices into the mesh (using Lawson's algorithm to maintain the Delaunay property) until all constraints on minimum angle and maximum triangle area are met (Fig. 9). Vertex insertion is governed by two rules.

- The *diametral circle* of a segment is the (unique) smallest circle that contains the segment. A segment is said to be *encroached* if a point lies within its diametral circle. Any encroached segment that arises is immediately split by inserting a vertex at its midpoint. The two resulting subsegments have smaller diametral circles, and may or may not be encroached themselves. See Fig. 10.
- The *circumcircle* of a triangle is the unique circle that passes through all three vertices of the triangle. A triangle is said to be *bad* if it has an angle too small or an area too large to satisfy the user's constraints. A bad triangle is split by inserting a vertex at its *circumcenter* (the center of its circumcircle); the Delaunay property guarantees that the triangle is eliminated (see Fig. 11). If the new vertex encroaches upon any segment, the vertex is deleted (reversing the insertion process) and all the segments it encroached upon are split.

Encroached segments are given priority over bad triangles. A queue of encroached segments and a queue of bad triangles are initialized at the beginning of

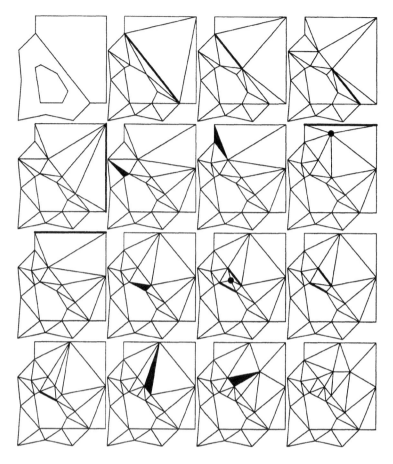

Fig. 12. Demonstration of the refinement stage. The first two images are the input PSLG and its constrained Delaunay triangulation. In each image, highlighted segments or triangles are about to be split, and highlighted vertices are about to be deleted. Note that the algorithm easily accommodates internal boundaries and holes.

the refinement stage and maintained throughout; every vertex insertion may add new members to either queue. The former queue rarely contains more than one segment except at the beginning of the refinement stage, when it may contain many.

The refinement stage is illustrated in Fig. 12. Ruppert [15] proves that this procedure halts for an angle constraint of up to 20.7°. In practice, the algorithm generally halts with an angle constraint of 33.8°, but often fails to terminate given an angle constraint of 33.9°. It would be interesting to discover why the cutoff falls there.

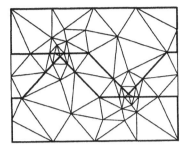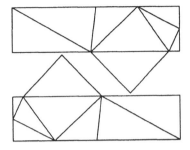

Fig. 13. Two variations of the Delaunay refinement algorithm with a 20° minimum angle. Left: Mesh created using segment splitting and late removal of triangles. This illustration includes external triangles, just prior to removal, to show why overrefinement occurs. Right: Mesh created using constrained Delaunay triangulation and early removal of triangles.

3.1 Selected Implementation Issues

Triangle removes extraneous triangles from holes and concavities before the refinement stage. This presents no problem for the refinement algorithm; the requirement that no segment be encroached and the Delaunay property together ensure that the circumcenter of every triangle lies within the mesh. (Roundoff error might perturb a circumcenter to just outside the mesh, but it is easy to identify the conflicting edge and treat it as encroached.) An advantage of removing triangles before refinement is that computation is not wasted refining triangles that will eventually be deleted.

A more important advantage is illustrated in Fig. 13. If extraneous triangles remain during the refinement stage, overrefinement can occur if very small features outside the object being meshed cause the creation of small triangles inside the mesh. Ruppert suggests solving this problem by using the constrained Delaunay triangulation, and ignoring interactions that take place outside the region being triangulated. Early removal of triangles provides a nearly effortless way to accomplish this effect. Segments that would normally be considered encroached are ignored (Fig. 13, right), because encroached segments are diagnosed by noticing that they occur opposite an obtuse angle in a triangle.

Another determinant of the number of triangles in the final mesh is the order in which bad triangles are split, especially when a strong angle constraint is used. Figure 14 demonstrates how sensitive the refinement algorithm is to the order. For this example with a 33° minimum angle, a heap of bad triangles indexed by their smallest angle confers a 35% reduction in mesh size over a first-in first-out queue. (This difference is typical for large meshes with a strong angle constraint, but thankfully disappears for small meshes and small constraints.) The discrepancy probably occurs because circumcenters of very bad triangles are likely to split more bad triangles than circumcenters of mildly bad triangles. Unfortunately, a heap is slow for large meshes, especially when small area constraints force all of the triangles into the heap. Delaunay refinement usually takes $\mathcal{O}(n)$

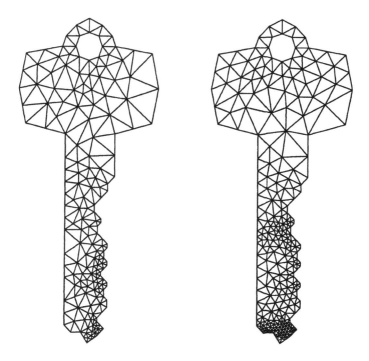

Fig. 14. Two meshes with a 33° minimum angle. The left mesh, with 290 triangles, was formed by always splitting the worst existing triangle. The right mesh, with 450 triangles, was formed by using a first-come first-split queue of bad triangles.

time in practice, but use of a heap increases the complexity to $\mathcal{O}(n \log n)$.

Triangle's solution, chosen experimentally, is to use 64 FIFO queues, each representing a different interval of angles. It is counterproductive (in practice) to order well-shaped triangles by their worst angle, so one queue is used for well-shaped but too-large triangles whose angles are all roughly larger than 39°. Triangles with smaller angles are partitioned among the remaining queues. When a bad triangle is chosen for splitting, it is taken from the "worst" nonempty queue. This method yields meshes comparable with those generated using a heap, but is only slightly slower than using a single queue. During the refinement phase, about 21,000 new vertices are generated per second on a DEC 3000/700. These vertices are inserted using the incremental Delaunay algorithm, but are inserted much more quickly than Table 1 would suggest because a triangle's circumcenter can be located quickly by starting the search at the triangle.

3.2 A Negative Result on Quality Triangulations

For any angle bound $\theta > 0$, there exists a PSLG $\mathcal{P}$ such that it is not possible to triangulate $\mathcal{P}$ without creating a new corner (not present in $\mathcal{P}$) having angle smaller than θ. Here, I discuss why this is true.

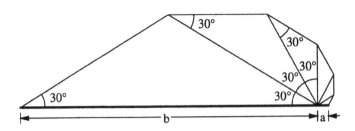

Fig. 15. In any triangulation with no angles smaller than 30°, the ratio b/a cannot exceed 27.

Ruppert's proof that his Delaunay refinement algorithm terminates makes use of the assumption that all interior angles are 90° or larger. This condition is often violated in practice, so he suggests handling small interior angles by surrounding each vertex of an acute angle with a ring of *shield edges*. As the negative result stated above suggests, there are PSLGs for which shield edges fail, and for which no construction can succeed. Fortunately, all such PSLGs I am aware of have an interior angle much smaller than θ, so failure is generally predictable.

The reasoning behind the result is as follows. Suppose a segment in a conforming triangulation has been split into two subsegments of lengths a and b, as illustrated in Fig. 15. Mitchell [13] proves that if the triangulation has no angles smaller than θ, then the ratio b/a has an upper bound of $(2\cos\theta)^{180°/\theta}$. (This bound is tight if $180°/\theta$ is an integer; Figure 15 offers an example where the bound is obtained.) Hence any bound on the smallest angle of a triangulation imposes a limit on the gradation of triangle sizes along a segment (or anywhere in the mesh).

A problem can arise if a small angle ϕ occurs at the intersection point o of two segments of a PSLG, as illustrated in Fig. 16 (top). The small angle cannot be improved, of course, but one does not wish to create any new small angles. Assume that one of the segments is split by a point p, which may be present in the input or may be inserted to help achieve the angle constraint elsewhere in the triangulation. The insertion of p forces part of the region between the two segments to be triangulated (Fig. 16, center), which can cause a new point q to be inserted on the segment containing p. Let $a = |\overline{pq}|$ and $b = |\overline{op}|$ as illustrated. If the angle bound is maintained, the length a cannot be large; the ratio a/b is bounded below

$$\frac{\sin\phi}{\sin\theta}\left(\cos(\theta+\phi) + \frac{\sin(\theta+\phi)}{\tan\theta}\right).$$

If the region above the segments is part of the interior of the PSLG, the fan effect demonstrated in Fig. 15 may necessitate the insertion of another vertex r between o and p (Fig. 16, bottom); this circumstance is unavoidable if the product of the bounds on b/a and a/b given above is less than one. For an angle constraint of $\theta = 30°$, this condition occurs when ϕ is about six tenths of a degree. Unfortunately, the new vertex r creates the same conditions as the

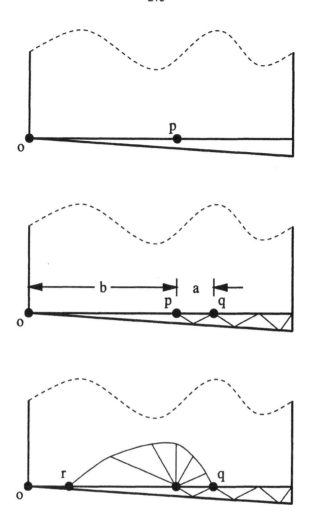

Fig. 16. Top: A difficult PSLG with a small interior angle ϕ. Center: The point p and the angle constraint necessitate the insertion of the point q. Bottom: The point q and the angle constraint necessitate the insertion of the point r. The process repeats eternally.

vertex p, but closer to o; the process will cascade, eternally creating smaller and smaller triangles in an attempt to satisfy the angle constraint. No algorithm can produce a finite triangulation of such a PSLG without violating the angle constraint. (It is amusing to consider whether the angle constraint can be met if one is allowed an infinite number of triangles.)

If some PSLGs do not have quality triangulations, what are the implications for shielding? Triangle implements a variant of shielding known as "modified segment splitting using concentric circular shells" (see Ruppert [15] for details), which is generally effective in practice for PSLGs that have small angles greater than 5°, and often for smaller angles. Shielding is useful even though it cannot solve all problems. On the other hand, the Delaunay refinement algorithm does not know to use careful arrangements of triangles as in Fig. 15 to manage small input angles, and therefore can fail to terminate even on PSLGs for which a quality triangulation exists. Hence, Triangle prints a warning message when angles smaller than five degrees appear between input segments. The smaller an angle is, and the greater the number of small angles in a PSLG, the less likely Triangle is to terminate. An interesting question for future work is how to determine when and where it is wise to weaken the angle constraint so that termination can be ensured.

This problem presents another motivation for removing triangles from holes and concavities prior to applying the Delaunay refinement algorithm. Holes with small angles might cause the algorithm to fail if triangles are not removed until after refinement. Concave objects can be particularly dastardly, because a very small angle may occur between a defining segment of the object and an edge of the convex hull. The user, unaware of the effect of the convex hull edge, would be mystified why the Delaunay refinement algorithm fails to terminate on what appears to be a simple PSLG. (In fact, this is how the issues described in this section first became evident to me.) Early removal of triangles from concavities avoids this problem.

4 Correct Adaptive Tests

The correctness of the incremental and divide-and-conquer algorithms depends on reliable *orientation* and *incircle* tests. The orientation test determines whether a point lies to the left of, to the right of, or on a line; it is used in many (perhaps most) geometric algorithms. The incircle test determines whether a point lies inside, outside, or on a circle. Inexact versions of these tests are vulnerable to roundoff error, and the wrong answers they produce can cause geometric algorithms to hang, crash, or produce incorrect output. Figure 17 demonstrates a real example of the failure of Triangle's divide-and-conquer algorithm.

The easiest solution to many of these robustness problems is to use software implementations of exact arithmetic, albeit often at great expense. It is common to hear reports of implementations being slowed by factors of ten or more as a consequence. The goal of improving the speed of correct geometric calculations has received much recent attention [4, 8, 1], but the most promising proposals take integer or rational inputs, often of limited precision. These methods do not appear to be usable if it is convenient or necessary to use ordinary floating-point inputs.

Triangle includes fast correct implementations of the orientation and incircle tests that take floating-point inputs. They owe their speed to two features. First, they employ new fast algorithms for arbitrary precision arithmetic that have a strong advantage over other software techniques in computations that manipulate values of extended but small precision. Second, they are adaptive; their running time depends on the degree of uncertainty of the result, and is usually

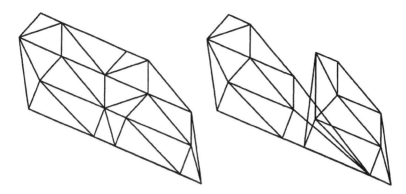

Fig. 17. Left: A Delaunay triangulation (two of the guitar's tuning screws). Right: An invalid triangulation created by Triangle with exact arithmetic disabled.

small. For instance, the adaptive orientation test is slow only if the points being tested are nearly or exactly collinear.

The orientation and incircle tests both work by computing the sign of a determinant. Fortune and Van Wyk [8] take advantage of the fact that only the sign is needed by using a *floating-point filter*: the determinant is first evaluated approximately, and only if forward error analysis indicates that the sign of the approximate result cannot be trusted does one use an exact test. Triangle's adaptive implementations carry this suggestion to its logical extreme by computing a sequence of successively more accurate approximations to the determinant, stopping only when the accuracy of the sign is assured. To reduce computation time, some of these approximations can reuse previous, less accurate computations. Shewchuk [16] presents details of the arbitrary precision arithmetic algorithms and the adaptivity scheme, and provides empirical evidence that multiple-stage adaptivity can significantly improve on two-stage adaptivity when difficult point sets are triangulated.

Using the adaptive tests, Triangle computes Delaunay triangulations, constrained Delaunay triangulations, and convex hulls exactly, roundoff error notwithstanding. Table 1 shows that the robust tests usually incur only a 10% to 30% overhead, though more time may be needed for points sets with many near-degeneracies. One exception is the divide-and-conquer algorithm with vertical cuts. Because this algorithm repeatedly merges tall, thinly separated triangulations, it performs many orientation tests on nearly-collinear points, and hence the robust version is much slower than the non-robust version. The variant that uses alternating cuts encounters nearly-collinear points less often; hence, its robust version suffers a smaller speed handicap, and its non-robust version is less likely to fail.

Of course, adaptive tests do not solve all robustness problems. Geometric computations that produce new vertices, including circumcenters and segment intersections, could be performed exactly in principle, but the results would have large bit complexity and would be inconvenient to manipulate and expensive to store. Worse, vertices of arbitrarily large bit complexity could eventually be produced in a cascading effect when the Delaunay refinement algorithm inserts circumcenters of triangles whose vertices were themselves circumcenters. Hence, it is infeasible to make the algorithm perfectly robust. Fortunately, the Delaunay refinement algorithm is naturally stable with regard to floating-point roundoff error. Problems arise only when triangles are refined to so small a size that it is no longer possible to construct a circumcenter that is distinct from its triangle's vertices.

I have not produced a robust version of the sweepline algorithm for a somewhat technical reason. The sweepline algorithm maintains a priority queue (normally implemented as a heap) containing two types of events: *site events*, where the sweepline passes over an input point, and *circle events*, where the sweepline reaches the top of a circle defined by three consecutive vertices on the boundary of the triangulation. Unfortunately, the y-coordinate of such a circle top is expensive to compute exactly, may be irrational, and has a somewhat complicated exact representation. A robust implementation must keep the events correctly ordered, and hence must replace the simple comparisons normally used to maintain a priority queue with a test that correctly compares two circle tops. Even a fast adaptive version of such a test would be so much slower than simple comparisons that event queue maintenance, which is a dominant cost of the sweepline algorithm, would become prohibitively expensive.

A   Additional Implementation Notes

The sweepline and incremental Delaunay triangulation implementations compared by Su and Drysdale [18] each use some variant of uniform bucketing to locate points. Bucketing yields fast implementations on uniform point sets, but is easily defeated; a small, dense cluster of points in a large, sparsely populated region may all fall into a single bucket. I have not used bucketing in Triangle, preferring algorithms that exhibit good performance with any distribution of input points. As a result, Triangle may be slower than necessary when triangulating uniformly distributed point sets, but will not exhibit asymptotically slower running times on difficult inputs.

Fortune's sweepline algorithm uses two nontrivial data structures in addition to the triangulation: a priority queue to store events, and a balanced tree data structure to store the sequence of edges on the boundary of the mesh. Fortune's own implementation, available from Netlib, uses bucketing to perform both these functions; hence, an $\mathcal{O}(n \log n)$ running time is not guaranteed, and Su and Drysdale [18] found that the original implementation exhibits $\mathcal{O}(n^{3/2})$ performance on uniform random point sets. By modifying Fortune's code to use a heap to store events, they obtained $\mathcal{O}(n \log n)$ running time and better performance on large point sets (having more than 50,000 points). However, bucketing outperforms a heap on small point sets.

Triangle's implementation uses a heap as well, and also uses a splay tree [17] to store mesh boundary edges, so that an $\mathcal{O}(n \log n)$ running time is attained, regardless of the distribution of points. Not all boundary edges are stored in the splay tree; when a new edge is created, it is inserted into the tree with probability 0.1. (The value 0.1 was chosen empirically to minimize the triangulation time for uniform random point sets.) At any time, the splay tree contains a random sample of roughly one tenth of the boundary edges. When the sweepline sweeps past an input point, the point must be located relative to the boundary edges; this point location involves a search in the splay tree, followed by a search on the boundary of the triangulation itself.

Splay trees adjust themselves so that frequently accessed items are near the top of the tree. Hence, a point set organized so that many new vertices appear at roughly the same location on the boundary of the mesh is likely to be triangulated quickly. This effect partly explains why Triangle's sweepline implementation triangulates points on the boundary of a circle more quickly than the other point sets, even though there are many more boundary edges in the cocircular point set and the splay tree grows to be much larger (containing $\mathcal{O}(n)$ boundary edges instead of $\mathcal{O}(\sqrt{n})$).

Triangle's incremental insertion algorithm for Delaunay triangulation uses the point location method proposed by Mücke, Saias, and Zhu [14]. Their *jump-and-walk* method chooses a random sample of $\mathcal{O}(n^{1/3})$ vertices from the mesh (where n is the number of nodes *currently* in the mesh), determines which of these vertices is closest to the query point, and walks through the mesh from the chosen vertex toward the query point until the triangle containing that point is found. Mücke et al. show that the resulting incremental algorithm takes expected $\mathcal{O}(n^{4/3})$ time on uniform random point sets. Table 1 appears to confirm this analysis. Triangle uses a sample size of $0.45n^{1/3}$; the coefficient was chosen empirically to minimize the triangulation time for uniform random point sets. Triangle also checks the previously inserted point, because in many practical point sets, any two consecutive points have a high likelihood of being near each other.

A more elaborate point location scheme such as that suggested by Guibas, Knuth, and Sharir [9] could be used (along with randomization of the insertion order) to obtain an expected $\mathcal{O}(n \log n)$ triangulation algorithm, but the data structure used for location is likely to take up as much memory as the triangulation itself, and unlikely to surpass the performance of the divide-and-conquer algorithm; hence, I do not intend to pursue it.

Note that all discussion in this paper applies to Triangle version 1.2; earlier versions lack the sweepline algorithm and many optimizations to the other algorithms.

Acknowledgments

Thanks to Dafna Talmor for comments on an early version of this paper. This research was supported in part by the Natural Sciences and Engineering Research Council of Canada under a 1967 Science and Engineering Scholarship and by the National Science Foundation under Grant CMS-9318163.

References

1. Francis Avnaim, Jean-Daniel Boissonnat, Olivier Devillers, Franco P. Preparata, and Mariette Yvinec. *Evaluating Signs of Determinants Using Single-Precision Arithmetic.* To appear in Algorithmica, 1995.
2. Marshall Bern and David Eppstein. *Mesh Generation and Optimal Triangulation.* Computing in Euclidean Geometry (Ding-Zhu Du and Frank Hwang, editors), Lecture Notes Series on Computing, volume 1, pages 23–90. World Scientific, Singapore, 1992.
3. L. Paul Chew. *Guaranteed-Quality Mesh Generation for Curved Surfaces.* Proceedings of the Ninth Annual Symposium on Computational Geometry, pages 274–280. Association for Computing Machinery, May 1993.
4. Kenneth L. Clarkson. *Safe and Effective Determinant Evaluation.* 33rd Annual Symposium on Foundations of Computer Science, pages 387–395. IEEE Computer Society Press, October 1992.
5. Rex A. Dwyer. *A Faster Divide-and-Conquer Algorithm for Constructing Delaunay Triangulations.* Algorithmica 2(2):137–151, 1987.
6. Steven Fortune. *A Sweepline Algorithm for Voronoï Diagrams.* Algorithmica 2(2):153–174, 1987.
7. ———. *Voronoï Diagrams and Delaunay Triangulations.* Computing in Euclidean Geometry (Ding-Zhu Du and Frank Hwang, editors), Lecture Notes Series on Computing, volume 1, pages 193–233. World Scientific, Singapore, 1992.
8. Steven Fortune and Christopher J. Van Wyk. *Efficient Exact Arithmetic for Computational Geometry.* Proceedings of the Ninth Annual Symposium on Computational Geometry, pages 163–172. Association for Computing Machinery, May 1993.
9. Leonidas J. Guibas, Donald E. Knuth, and Micha Sharir. *Randomized Incremental Construction of Delaunay and Voronoï Diagrams.* Algorithmica 7(4):381–413, 1992.
10. Leonidas J. Guibas and Jorge Stolfi. *Primitives for the Manipulation of General Subdivisions and the Computation of Voronoï Diagrams.* ACM Transactions on Graphics 4(2):74–123, April 1985.
11. C. L. Lawson. *Software for C^1 Surface Interpolation.* Mathematical Software III (John R. Rice, editor), pages 161–194. Academic Press, New York, 1977.
12. D. T. Lee and B. J. Schachter. *Two Algorithms for Constructing a Delaunay Triangulation.* International Journal of Computer and Information Sciences 9(3):219–242, 1980.
13. Scott A. Mitchell. *Cardinality Bounds for Triangulations with Bounded Minimum Angle.* Sixth Canadian Conference on Computational Geometry, 1994.
14. Ernst P. Mücke, Isaac Saias, and Binhai Zhu. *Fast Randomized Point Location Without Preprocessing in Two- and Three-dimensional Delaunay Triangulations.* Proceedings of the Twelfth Annual Symposium on Computational Geometry, pages 274–283. Association for Computing Machinery, May 1996.
15. Jim Ruppert. *A Delaunay Refinement Algorithm for Quality 2-Dimensional Mesh Generation.* Journal of Algorithms 18(3):548–585, May 1995.
16. Jonathan Richard Shewchuk. *Robust Adaptive Floating-Point Geometric Predicates.* Proceedings of the Twelfth Annual Symposium on Computational Geometry, pages 141–150. Association for Computing Machinery, May 1996.
17. Daniel Dominic Sleator and Robert Endre Tarjan. *Self-Adjusting Binary Search Trees.* Journal of the Association for Computing Machinery 32(3):652–686, July 1985.

18. Peter Su and Robert L. Scot Drysdale. *A Comparison of Sequential Delaunay Triangulation Algorithms*. Proceedings of the Eleventh Annual Symposium on Computational Geometry, pages 61–70. Association for Computing Machinery, June 1995.

Author's Index

Springer-Verlag
and the Environment

We at Springer-Verlag firmly believe that an international science publisher has a special obligation to the environment, and our corporate policies consistently reflect this conviction.

We also expect our business partners – paper mills, printers, packaging manufacturers, etc. – to commit themselves to using environmentally friendly materials and production processes.

The paper in this book is made from low- or no-chlorine pulp and is acid free, in conformance with international standards for paper permanency.

Lecture Notes in Computer Science

For information about Vols. 1–1081

please contact your bookseller or Springer-Verlag